D0998984

# FREE Test Taking Tips DVD Offer

To help us better serve you, we have developed a Test Taking Tips DVD that we would like to give you for FREE. **This DVD covers world-class test taking tips that you can use to be even more successful when you are taking your test.**

All that we ask is that you email us your feedback about your study guide. Please let us know what you thought about it – whether that is good, bad or indifferent.

To get your **FREE Test Taking Tips DVD**, email freedvd@studyguideteam.com with "FREE DVD" in the subject line and the following information in the body of the email:

a. The title of your study guide.

b. Your product rating on a scale of 1-5, with 5 being the highest rating.

c. Your feedback about the study guide. What did you think of it?

d. Your full name and shipping address to send your free DVD.

If you have any questions or concerns, please don't hesitate to contact us at freedvd@studyguideteam.com.

Thanks again!

# SAT Subject Test Math Level 2 Study Guide

## SAT Math 2 Subject Test Prep and Practice Exam Questions [4th Edition]

TPB Publishing

Copyright © 2020 by TPB Publishing

All rights reserved. No part of this publication may be reproduced, distributed, or transmitted in any form or by any means, including photocopying, recording, or other electronic or mechanical methods, without the prior written permission of the publisher, except in the case of brief quotations embodied in critical reviews and certain other noncommercial uses permitted by copyright law.

Written and edited by TPB Publishing.

TPB Publishing is not associated with or endorsed by any official testing organization. TPB Publishing is a publisher of unofficial educational products. All test and organization names are trademarks of their respective owners. Content in this book is included for utilitarian purposes only and does not constitute an endorsement by TPB Publishing of any particular point of view.

Interested in buying more than 10 copies of our product? Contact us about bulk discounts: bulkorders@studyguideteam.com

ISBN 13: 9781628459890
ISBN 10: 1628459891

# Table of Contents

# Quick Overview

As you draw closer to taking your exam, effective preparation becomes more and more important. Thankfully, you have this study guide to help you get ready. Use this guide to help keep your studying on track and refer to it often.

This study guide contains several key sections that will help you be successful on your exam. The guide contains tips for what you should do the night before and the day of the test. Also included are test-taking tips. Knowing the right information is not always enough. Many well-prepared test takers struggle with exams. These tips will help equip you to accurately read, assess, and answer test questions.

A large part of the guide is devoted to showing you what content to expect on the exam and to helping you better understand that content. In this guide are practice test questions so that you can see how well you have grasped the content. Then, answer explanations are provided so that you can understand why you missed certain questions.

Don't try to cram the night before you take your exam. This is not a wise strategy for a few reasons. First, your retention of the information will be low. Your time would be better used by reviewing information you already know rather than trying to learn a lot of new information. Second, you will likely become stressed as you try to gain a large amount of knowledge in a short amount of time. Third, you will be depriving yourself of sleep. So be sure to go to bed at a reasonable time the night before. Being well-rested helps you focus and remain calm.

Be sure to eat a substantial breakfast the morning of the exam. If you are taking the exam in the afternoon, be sure to have a good lunch as well. Being hungry is distracting and can make it difficult to focus. You have hopefully spent lots of time preparing for the exam. Don't let an empty stomach get in the way of success!

When travelling to the testing center, leave earlier than needed. That way, you have a buffer in case you experience any delays. This will help you remain calm and will keep you from missing your appointment time at the testing center.

Be sure to pace yourself during the exam. Don't try to rush through the exam. There is no need to risk performing poorly on the exam just so you can leave the testing center early. Allow yourself to use all of the allotted time if needed.

Remain positive while taking the exam even if you feel like you are performing poorly. Thinking about the content you should have mastered will not help you perform better on the exam.

Once the exam is complete, take some time to relax. Even if you feel that you need to take the exam again, you will be well served by some down time before you begin studying again. It's often easier to convince yourself to study if you know that it will come with a reward!

# Test-Taking Strategies

## 1. Predicting the Answer

When you feel confident in your preparation for a multiple-choice test, try predicting the answer before reading the answer choices. This is especially useful on questions that test objective factual knowledge. By predicting the answer before reading the available choices, you eliminate the possibility that you will be distracted or led astray by an incorrect answer choice. You will feel more confident in your selection if you read the question, predict the answer, and then find your prediction among the answer choices. After using this strategy, be sure to still read all of the answer choices carefully and completely. If you feel unprepared, you should not attempt to predict the answers. This would be a waste of time and an opportunity for your mind to wander in the wrong direction.

## 2. Reading the Whole Question

Too often, test takers scan a multiple-choice question, recognize a few familiar words, and immediately jump to the answer choices. Test authors are aware of this common impatience, and they will sometimes prey upon it. For instance, a test author might subtly turn the question into a negative, or he or she might redirect the focus of the question right at the end. The only way to avoid falling into these traps is to read the entirety of the question carefully before reading the answer choices.

## 3. Looking for Wrong Answers

Long and complicated multiple-choice questions can be intimidating. One way to simplify a difficult multiple-choice question is to eliminate all of the answer choices that are clearly wrong. In most sets of answers, there will be at least one selection that can be dismissed right away. If the test is administered on paper, the test taker could draw a line through it to indicate that it may be ignored; otherwise, the test taker will have to perform this operation mentally or on scratch paper. In either case, once the obviously incorrect answers have been eliminated, the remaining choices may be considered. Sometimes identifying the clearly wrong answers will give the test taker some information about the correct answer. For instance, if one of the remaining answer choices is a direct opposite of one of the eliminated answer choices, it may well be the correct answer. The opposite of obviously wrong is obviously right! Of course, this is not always the case. Some answers are obviously incorrect simply because they are irrelevant to the question being asked. Still, identifying and eliminating some incorrect answer choices is a good way to simplify a multiple-choice question.

## 4. Don't Overanalyze

Anxious test takers often overanalyze questions. When you are nervous, your brain will often run wild, causing you to make associations and discover clues that don't actually exist. If you feel that this may be a problem for you, do whatever you can to slow down during the test. Try taking a deep breath or counting to ten. As you read and consider the question, restrict yourself to the particular words used by the author. Avoid thought tangents about what the author *really* meant, or what he or she was *trying* to say. The only things that matter on a multiple-choice test are the words that are actually in the question. You must avoid reading too much into a multiple-choice question, or supposing that the writer meant something other than what he or she wrote.

## 5. No Need for Panic

It is wise to learn as many strategies as possible before taking a multiple-choice test, but it is likely that you will come across a few questions for which you simply don't know the answer. In this situation, avoid panicking. Because most multiple-choice tests include dozens of questions, the relative value of a single wrong answer is small. As much as possible, you should compartmentalize each question on a multiple-choice test. In other words, you should not allow your feelings about one question to affect your success on the others. When you find a question that you either don't understand or don't know how to answer, just take a deep breath and do your best. Read the entire question slowly and carefully. Try rephrasing the question a couple of different ways. Then, read all of the answer choices carefully. After eliminating obviously wrong answers, make a selection and move on to the next question.

## 6. Confusing Answer Choices

When working on a difficult multiple-choice question, there may be a tendency to focus on the answer choices that are the easiest to understand. Many people, whether consciously or not, gravitate to the answer choices that require the least concentration, knowledge, and memory. This is a mistake. When you come across an answer choice that is confusing, you should give it extra attention. A question might be confusing because you do not know the subject matter to which it refers. If this is the case, don't eliminate the answer before you have affirmatively settled on another. When you come across an answer choice of this type, set it aside as you look at the remaining choices. If you can confidently assert that one of the other choices is correct, you can leave the confusing answer aside. Otherwise, you will need to take a moment to try to better understand the confusing answer choice. Rephrasing is one way to tease out the sense of a confusing answer choice.

## 7. Your First Instinct

Many people struggle with multiple-choice tests because they overthink the questions. If you have studied sufficiently for the test, you should be prepared to trust your first instinct once you have carefully and completely read the question and all of the answer choices. There is a great deal of research suggesting that the mind can come to the correct conclusion very quickly once it has obtained all of the relevant information. At times, it may seem to you as if your intuition is working faster even than your reasoning mind. This may in fact be true. The knowledge you obtain while studying may be retrieved from your subconscious before you have a chance to work out the associations that support it. Verify your instinct by working out the reasons that it should be trusted.

## 8. Key Words

Many test takers struggle with multiple-choice questions because they have poor reading comprehension skills. Quickly reading and understanding a multiple-choice question requires a mixture of skill and experience. To help with this, try jotting down a few key words and phrases on a piece of scrap paper. Doing this concentrates the process of reading and forces the mind to weigh the relative importance of the question's parts. In selecting words and phrases to write down, the test taker thinks about the question more deeply and carefully. This is especially true for multiple-choice questions that are preceded by a long prompt.

## 9. Subtle Negatives

One of the oldest tricks in the multiple-choice test writer's book is to subtly reverse the meaning of a question with a word like *not* or *except*. If you are not paying attention to each word in the question, you can easily be led astray by this trick. For instance, a common question format is, "Which of the following is...?" Obviously, if the question instead is, "Which of the following is not...?," then the answer will be quite different. Even worse, the test makers are aware of the potential for this mistake and will include one answer choice that would be correct if the question were not negated or reversed. A test taker who misses the reversal will find what he or she believes to be a correct answer and will be so confident that he or she will fail to reread the question and discover the original error. The only way to avoid this is to practice a wide variety of multiple-choice questions and to pay close attention to each and every word.

## 10. Reading Every Answer Choice

It may seem obvious, but you should always read every one of the answer choices! Too many test takers fall into the habit of scanning the question and assuming that they understand the question because they recognize a few key words. From there, they pick the first answer choice that answers the question they believe they have read. Test takers who read all of the answer choices might discover that one of the latter answer choices is actually *more* correct. Moreover, reading all of the answer choices can remind you of facts related to the question that can help you arrive at the correct answer. Sometimes, a misstatement or incorrect detail in one of the latter answer choices will trigger your memory of the subject and will enable you to find the right answer. Failing to read all of the answer choices is like not reading all of the items on a restaurant menu: you might miss out on the perfect choice.

## 11. Spot the Hedges

One of the keys to success on multiple-choice tests is paying close attention to every word. This is never truer than with words like almost, most, some, and sometimes. These words are called "hedges" because they indicate that a statement is not totally true or not true in every place and time. An absolute statement will contain no hedges, but in many subjects, the answers are not always straightforward or absolute. There are always exceptions to the rules in these subjects. For this reason, you should favor those multiple-choice questions that contain hedging language. The presence of qualifying words indicates that the author is taking special care with his or her words, which is certainly important when composing the right answer. After all, there are many ways to be wrong, but there is only one way to be right! For this reason, it is wise to avoid answers that are absolute when taking a multiple-choice test. An absolute answer is one that says things are either all one way or all another. They often include words like *every*, *always*, *best*, and *never*. If you are taking a multiple-choice test in a subject that doesn't lend itself to absolute answers, be on your guard if you see any of these words.

## 12. Long Answers

In many subject areas, the answers are not simple. As already mentioned, the right answer often requires hedges. Another common feature of the answers to a complex or subjective question are qualifying clauses, which are groups of words that subtly modify the meaning of the sentence. If the question or answer choice describes a rule to which there are exceptions or the subject matter is complicated, ambiguous, or confusing, the correct answer will require many words in order to be expressed clearly and accurately. In essence, you should not be deterred by answer choices that seem excessively long. Oftentimes, the author of the text will not be able to write the correct answer without offering some qualifications and modifications. Your job is to read the answer choices thoroughly and completely and to select the one that most accurately and precisely answers the question.

## 13. Restating to Understand

Sometimes, a question on a multiple-choice test is difficult not because of what it asks but because of how it is written. If this is the case, restate the question or answer choice in different words. This process serves a couple of important purposes. First, it forces you to concentrate on the core of the question. In order to rephrase the question accurately, you have to understand it well. Rephrasing the question will concentrate your mind on the key words and ideas. Second, it will present the information to your mind in a fresh way. This process may trigger your memory and render some useful scrap of information picked up while studying.

## 14. True Statements

Sometimes an answer choice will be true in itself, but it does not answer the question. This is one of the main reasons why it is essential to read the question carefully and completely before proceeding to the answer choices. Too often, test takers skip ahead to the answer choices and look for true statements. Having found one of these, they are content to select it without reference to the question above. Obviously, this provides an easy way for test makers to play tricks. The savvy test taker will always read the entire question before turning to the answer choices. Then, having settled on a correct answer choice, he or she will refer to the original question and ensure that the selected answer is relevant. The mistake of choosing a correct-but-irrelevant answer choice is especially common on questions related to specific pieces of objective knowledge. A prepared test taker will have a wealth of factual knowledge at his or her disposal, and should not be careless in its application.

## 15. No Patterns

One of the more dangerous ideas that circulates about multiple-choice tests is that the correct answers tend to fall into patterns. These erroneous ideas range from a belief that B and C are the most common right answers, to the idea that an unprepared test-taker should answer "A-B-A-C-A-D-A-B-A." It cannot be emphasized enough that pattern-seeking of this type is exactly the WRONG way to approach a multiple-choice test. To begin with, it is highly unlikely that the test maker will plot the correct answers according to some predetermined pattern. The questions are scrambled and delivered in a random order. Furthermore, even if the test maker was following a pattern in the assignation of correct answers, there is no reason why the test taker would know which pattern he or she was using. Any attempt to discern a pattern in the answer choices is a waste of time and a distraction from the real work of taking the test. A test taker would be much better served by extra preparation before the test than by reliance on a pattern in the answers.

# FREE DVD OFFER

Don't forget that doing well on your exam includes both understanding the test content and understanding how to use what you know to do well on the test. We offer a completely FREE Test Taking Tips DVD that covers world class test taking tips that you can use to be even more successful when you are taking your test.

All that we ask is that you email us your feedback about your study guide. To get your **FREE Test Taking Tips DVD**, email freedvd@studyguideteam.com with "FREE DVD" in the subject line and the following information in the body of the email:

- The title of your study guide.
- Your product rating on a scale of 1-5, with 5 being the highest rating.
- Your feedback about the study guide. What did you think of it?
- Your full name and shipping address to send your free DVD.

# Introduction

## Function of the Test

The Mathematics Level 2 Subject Test is a test that examines your knowledge of geometry, algebra, trigonometry, and precalculus in order to showcase your abilities through college admissions. Some colleges use Mathematics 2 to place students in appropriate courses in college, allowing them to pass over the basics into more complex courses. Note that in New York State, some people may use SAT Subject Test scores as a substitute for a Regents examination score. The Mathematics Level 2 Subject Test examines your knowledge of mathematics within a two-year period of algebra and a one-year period of geometry, with the additional familiarity of trigonometry or precalculus.

## Test Administration

To find out where the Mathematics Level 2 Subject Test is given, go to the website at collegereadiness.collegeboard.org. There you can find exam dates for the current year as well as up to three years in advance. The dates are listed by subject and give a deadline to register for each date. Currently, there are six dates listed each year for taking the subject tests, all of which offer the Mathematics 2 subject test. The dates for the exams are offered in August, October, November, December, May, and June. Note that you cannot take the SAT and an SAT Subject Test on the same date. The College Board website also has a test date finder, where you enter in the test date and your country, state, and city. The results bring up testing centers in your area or the area(s) nearest you.

## Test Format

The Mathematics Level 2 Subject test is sixty minutes long and has fifty multiple choice questions. The skills that are needed in order to be successful on the exam are numbers and operations; algebra and functions; geometry and measurement; and statistics, probability, and data analysis. The test sections are divided according to percentages in the table below:

| Content | Approximate % of Test |
|---|---|
| Number and operations | 10 to 14% |
| Algebra and functions | 48 to 52% |
| Geometry and measurement<br>(Coordinate)<br>(Three-dimensional)<br>(Trigonometry) | 48 to 52%<br>(10 to 14%)<br>(4 to 6%)<br>(12 to 16%) |
| Data analysis, statistics and probability | 8 to 12% |

Note that the difference between Mathematics I and Mathematics II is that Mathematics II is for test takers who are adept in the subjects of trigonometry or precalculus and know how to use a graphing calculator. In order to decide between the two tests, take a look at which subjects you are recently familiar with and what college programs you are most interested in.

Test takers are allowed to bring calculators for the exam day. Keep in mind that many of the answer choices on the Mathematics 2 exam are rounded, so the calculator answer may not be the exact answer choice you see on the exam. Remember to read each problem carefully and to not jump at the very first number that matches an answer choice. Follow the question through to the end.

## Scoring

In 2016, 145,140 students took the Mathematics Level 2 Subject Test. Out of a 200 to 800-point scale, the mean score was 690. The standard deviation was 101; this calculates the spread of grades around the average score. A "good score" for the Mathematics Level 2 Subject Test depends on the college you are applying to. Although many colleges are flexible and look at SAT scores along with other factors for admissions, highly selective schools prefer to see a score in the 700s or above. Note that you do not have to take SAT Subject Tests for admission to most colleges; even many highly selective schools recommend but do not require the exam.

## Recent Developments

The only new developments for the subject tests are changes to the testing calendar. The subject tests are offered on certain Saturdays in August, October, November, December, May, and June. For students who have to miss the Saturday test for religious reasons, the test is also offered on Sundays. Exam dates are no longer offered in January. Testing is available internationally, and subject tests will be offered outside of the U.S. and U.S. territories in October, November, December, May, and June.

# Number and Operations

## Basic Operations of Arithmetic

There are four different basic operations used with numbers: addition, subtraction, multiplication, and division.

- Addition takes two numbers and combines them into a total called the **sum**. The sum is the total when combining two collections into one. If there are 5 things in one collection and 3 in another, then after combining them, there is a total of $5 + 3 = 8$. Note the order does not matter when adding numbers. For example, $3 + 5 = 8$.

- Subtraction is the opposite (or "inverse") operation to addition. Whereas addition combines two quantities together, subtraction takes one quantity away from another. For example, if there are 20 gallons of fuel and 5 are removed, that gives $20 - 5 = 15$ gallons remaining. Note that for subtraction, the order does matter because it makes a difference which quantity is being removed from which.

- Multiplication is repeated addition. $3 \times 4$ can be thought of as putting together 3 sets of items, each set containing 4 items. The total is 12 items. Another way to think of this is to think of each number as the length of one side of a rectangle. If a rectangle is covered in tiles with 3 columns of 4 tiles each, then there are 12 tiles in total. From this, one can see that the answer is the same if the rectangle has 4 rows of 3 tiles each: $4 \times 3 = 12$. By expanding this reasoning, the order the numbers are multiplied does not matter.

- Division is the opposite of multiplication. It means taking one quantity and dividing it into sets the size of the second quantity. If there are 16 sandwiches to be distributed to 4 people, then each person gets $16 \div 4 = 4$ sandwiches. As with subtraction, the order in which the numbers appear does matter for division.

### Addition

Addition is the combination of two numbers so their quantities are added together cumulatively. The sign for an addition operation is the + symbol. For example, 9 + 6 = 15. The 9 and 6 combine to achieve a cumulative value, called a **sum**.

Addition holds the **commutative property**, which means that numbers in an addition equation can be switched without altering the result. The formula for the commutative property is a + b = b + a. Let's look at a few examples to see how the commutative property works:

$$7 = 3 + 4 = 4 + 3 = 7$$

$$20 = 12 + 8 = 8 + 12 = 20$$

Addition also holds the **associative property**, which means that the grouping of numbers doesn't matter in an addition problem. In other words, the presence or absence of parentheses is irrelevant. The formula for the associative property is $(a + b) + c = a + (b + c)$.

Here are some examples of the associative property at work:

$$30 = (6 + 14) + 10 = 6 + (14 + 10) = 30$$

$$35 = 8 + (2 + 25) = (8 + 2) + 25 = 35$$

## Subtraction

Subtraction is taking away one number from another, so their quantities are reduced. The sign designating a subtraction operation is the – symbol, and the result is called the **difference**. For example, $9 - 6 = 3$. The number *6* detracts from the number *9* to reach the difference *3*.

Unlike addition, subtraction follows neither the commutative nor associative properties. The order and grouping in subtraction impact the result.

$$15 = 22 - 7 \neq 7 - 22 = -15$$

$$3 = (10 - 5) - 2 \neq 10 - (5 - 2) = 7$$

When working through subtraction problems involving larger numbers, it's necessary to regroup the numbers. Let's work through a practice problem using regrouping:

$$\begin{array}{r} 3\ 2\ 5 \\ -\ \ 7\ 7 \\ \hline \end{array}$$

Here, it is clear that the ones and tens columns for 77 are greater than the ones and tens columns for 325. To subtract this number, borrow from the tens and hundreds columns. When borrowing from a column, subtracting 1 from the lender column will add 10 to the borrower column:

$$\begin{array}{rrr} 3-1 & 10+2-1 & 10+5 \\ - & 7 & 7 \\ \hline \end{array} = \begin{array}{rrr} 2 & 11 & 15 \\ - & 7 & 7 \\ \hline 2 & 4 & 8 \end{array}$$

After ensuring that each digit in the top row is greater than the digit in the corresponding bottom row, subtraction can proceed as normal, and the answer is found to be 248.

## Multiplication

Multiplication involves adding together multiple copies of a number. It is indicated by an × symbol or a number immediately outside of a parenthesis. For example:

$$5(8-2)$$

The two numbers being multiplied together are called **factors,** and their result is called a **product**. For example, $9 \times 6 = 54$. This can be shown alternatively by expansion of either the 9 or the 6:

$$9 \times 6 = 9 + 9 + 9 + 9 + 9 + 9 = 54$$

$$9 \times 6 = 6 + 6 + 6 + 6 + 6 + 6 + 6 + 6 + 6 = 54$$

Like addition, multiplication holds the commutative and associative properties:

$$115 = 23 \times 5 = 5 \times 23 = 115$$

$$84 = 3 \times (7 \times 4) = (3 \times 7) \times 4 = 84$$

Multiplication also follows the **distributive property**, which allows the multiplication to be distributed through parentheses. The formula for distribution is $a \times (b + c) = ab + ac$. This is clear after the examples:

$$45 = 5 \times 9 = 5(3 + 6) = (5 \times 3) + (5 \times 6) = 15 + 30 = 45$$

$$20 = 4 \times 5 = 4(10 - 5) = (4 \times 10) - (4 \times 5) = 40 - 20 = 20$$

Multiplication becomes slightly more complicated when multiplying numbers with decimals. The easiest way to answer these problems is to ignore the decimals and multiply as if they were whole numbers. After multiplying the factors, place a decimal in the product. The placement of the decimal is determined by taking the cumulative number of decimal places in the factors.

For example:

$$\begin{array}{r} 0.7 \\ \times 3 \\ \hline 2.1 \end{array} \qquad \begin{array}{r} 2.6 \\ \times\ 4.2 \\ \hline 10.92 \end{array} \qquad \begin{array}{r} 1.5 \\ \times 6.4 \\ \hline 9.60 \end{array}$$

Let's tackle the first example. First, ignore the decimal and multiply the numbers as though they were whole numbers to arrive at a product: 21. Second, count the number of digits that follow a decimal (one). Finally, move the decimal place that many positions to the left, as the factors have only one decimal place. The second example works the same way, except that there are two total decimal places in the factors, so the product's decimal is moved two places over. In the third example, the decimal should be moved over two digits, but the digit zero is no longer needed, so it is erased and the final answer is 9.6.

## Division

Division and multiplication are inverses of each other in the same way that addition and subtraction are opposites. The signs designating a division operation are the ÷ and / symbols. In division, the second number divides into the first.

The number before the division sign is called the **dividend** or, if expressed as a fraction, the **numerator**. For example, in $a \div b$, *a* is the dividend, while in $\frac{a}{b}$, *a* is the numerator.

The number after the division sign is called the **divisor** or, if expressed as a fraction, the **denominator**. For example, in $a \div b$, *b* is the divisor, while in $\frac{a}{b}$, *b* is the denominator.

Like subtraction, division doesn't follow the commutative property, as it matters which number comes before the division sign, and division doesn't follow the associative or distributive properties for the same reason.

For example:

$$\frac{3}{2} = 9 \div 6 \neq 6 \div 9 = \frac{2}{3}$$

$$2 = 10 \div 5 = (30 \div 3) \div 5 \neq 30 \div (3 \div 5) = 30 \div \frac{3}{5} = 50$$

$$25 = 20 + 5 = (40 \div 2) + (40 \div 8) \neq 40 \div (2 + 8) = 40 \div 10 = 4$$

If a divisor doesn't divide into a dividend an integer number of times, whatever is left over is termed the **remainder**. The remainder can be further divided out into decimal form by using long division; however, this doesn't always give a quotient with a finite number of decimal places, so the remainder can also be expressed as a fraction over the original divisor.

Division with decimals is similar to multiplication with decimals in that when dividing a decimal by a whole number, ignore the decimal and divide as if it were a whole number.

Upon finding the answer, or quotient, place the decimal at the decimal place equal to that in the dividend.

$$15.75 \div 3 = 5.25$$

When the divisor is a decimal number, multiply both the divisor and dividend by 10. Repeat this until the divisor is a whole number, then complete the division operation as described above.

$$17.5 \div 2.5 = 175 \div 25 = 7$$

## Order of Operations
When working with complicated expressions, parentheses are used to indicate in which order to perform operations. However, to avoid having too many parentheses in an expression, here are some basic rules concerning the proper order to perform operations when not otherwise specified.

1. Parentheses: always perform operations inside parentheses first, regardless of what those operations are

2. Exponents

3. Multiplication and Division

4. Addition and Subtraction

For #3 & #4, work these from left to right. So, if there a subtraction problem and then an addition problem, the subtraction problem will be worked first.

Note that multiplication and division are performed from left to right as they appear in the expression or equation. Addition and subtraction also are performed from left to right as they appear.

As an aid to memorizing this, some students like to use the mnemonic PEMDAS. Furthermore, this acronym can be associated with a mnemonic phrase such as "Pirates Eat Many Donuts At Sea."

## Properties of Operations
Properties of operations exist that make calculations easier and solve problems for missing values. The following table summarizes commonly used properties of real numbers.

| **Property** | **Addition** | **Multiplication** |
|---|---|---|
| Commutative | $a + b = b + a$ | $a \times b = b \times a$ |
| Associative | $(a + b) + c = a + (b + c)$ | $(a \times b) \times c = a \times (bc)$ |
| Identity | $a + 0 = a;\ 0 + a = a$ | $a \times 1 = a;\ 1 \times a = a$ |
| Inverse | $a + (-a) = 0$ | $a \times \frac{1}{a} = 1;\ a \neq 0$ |
| Distributive | $a(b + c) = ab + ac$ | |

The cumulative property of addition states that the order in which numbers are added does not change the sum. Similarly, the commutative property of multiplication states that the order in which numbers are multiplied does not change the product. The associative property of addition and multiplication state that the grouping of numbers being added or multiplied does not change the sum or product, respectively. The commutative and associative properties are useful for performing calculations. For example, $(47 + 25) + 3$ is equivalent to $(47 + 3) + 25$, which is easier to calculate.

The identity property of addition states that adding zero to any number does not change its value. The identity property of multiplication states that multiplying a number by 1 does not change its value. The inverse property of addition states that the sum of a number and its opposite equals zero. Opposites are numbers that are the same with different signs (ex. 5 and -5; $-\frac{1}{2}$ and $\frac{1}{2}$). The inverse property of multiplication states that the product of a number (other than 0) and its reciprocal equals 1. Reciprocal numbers have numerators and denominators that are inverted (ex. $\frac{2}{5}$ and $\frac{5}{2}$). Inverse properties are useful for canceling quantities to find missing values (see algebra content). For example, $a + 7 = 12$ is solved by adding the inverse of 7 (which is -7) to both sides in order to isolate $a$.

The distributive property states that multiplying a sum (or difference) by a number produces the same result as multiplying each value in the sum (or difference) by the number and adding (or subtracting) the products. Consider the following scenario: You are buying three tickets for a baseball game. Each ticket costs $18. You are also charged a fee of $2 per ticket for purchasing the tickets online. The cost is calculated: $3 \times 18 + 3 \times 2$. Using the distributive property, the cost can also be calculated $3(18 + 2)$.

## Using Operations in Math and Real-World Problems

As mentioned, addition and subtraction are "inverse operations." Adding a number and then subtracting the same number will cancel each other out. This results in the original number, and vice versa. For example, $8 + 7 - 7 = 8$ and $137 - 100 + 100 = 137$.

Multiplication and division are also **inverse operations**. So, multiplying by a number and then dividing by the same number results in the original number. For example, $8 \times 2 \div 2 = 8$ and $12 \div 4 \times 4 = 12$. Inverse operations are used to work backwards to solve problems. In the case that 7 and a number add to 18, the inverse operation of subtraction is used to find the unknown value $(18 - 7 = 11)$. If a school's entire 4$^{th}$ grade was divided evenly into 3 classes each with 22 students, the inverse operation of multiplication is used to determine the total students in the grade $(22 \times 3 = 66)$. More scenarios involving inverse operations are listed in the tables below.

Word problems take concepts you learned in the classroom and turn them into real-life situations. Some parts of the problem are known and at least one part is unknown. There are three types of instances in which something can be unknown: the starting point, the change, or the final result. Any of these can be missing from the information they give you.

For an addition problem, the change is the quantity of a new amount added to the starting point.

For a subtraction problem, the change is the quantity taken away from the starting point.

Regarding addition, the given equation is $3 + 7 = 10$.

The number 3 is the starting point. 7 is the change, and 10 is the result from adding a new amount to the starting point. Different word problems can arise from this same equation, depending on which value is the unknown. For example, here are three problems:

- If a boy had 3 pencils and was given 7 more, how many would he have in total?
- If a boy had 3 pencils and a girl gave him more so that he had 10 in total, how did she give to him?
- A boy was given 7 pencils so that he had 10 in total. How many did he start with?

All three problems involve the same equation. Finding out which part of the equation is missing is the key to solving each word problem. The missing answers would be 10, 7, and 3, respectively.

In terms of subtraction, the same three scenarios can occur. Imagine the given equation is $6 - 4 = 2$.

The number 6 is the starting point, 4 is the change, and 2 is the new amount that is the result from taking away an amount from the starting point. Again, different types of word problems can arise from this equation. For example, here are three possible problems:

- If a girl had 6 quarters and 2 were taken away, how many would be left over?
- If a girl had 6 quarters, purchased a pencil, and had 2 quarters left over, how many did she pay with?
- If a girl paid for a pencil with 4 quarters and had 2 quarters left over, how many did she start with?

The three question types follow the structure of the addition word problems. Finding out whether the starting point, the change, or the final result is missing is the goal in solving the problem. The missing answers would be 2, 4, and 6, respectively.

The three addition problems and the three subtraction word problems can be solved by using a picture, a number line, or an algebraic equation. If an equation is used, a question mark can be used to show the number we don't know. For example, $6 - 4 = ?$ can be written to show that the missing value is the result. Using equation form shows us what part of the addition or subtraction problem is missing.

Key words within a multiplication problem involve *times, product, doubled*, and *tripled.* Key words within a division problem involve *split, quotient, divided, shared, groups*, and *half*. Like addition and subtraction, multiplication and division problems also have three different types of missing values.

Multiplication consists of a certain number of groups, with the same number of items within each group, and the total amount within all groups. Therefore, each one of these amounts can be the missing value.

For example, the given equation is $5 \times 3 = 15$.

5 and 3 are interchangeable, so either amount can be the number of groups or the number of items within each group. 15 is the total number of items. Again, different types of word problems can arise from this equation. For example, here are three problems:

- If a classroom is serving 5 different types of apples for lunch and has 3 apples of each type, how many total apples are there to give to the students?
- If a classroom has 15 apples with 5 different types, how many of each type are there?
- If a classroom has 15 apples total with 3 of each type, how many types are there to choose from?

Each question involves using the same equation to solve. It is important to decide which part of the equation is the missing value. The answers to the problems are 15, 3, and 5, respectively.

Similar to multiplication, division problems involve a total amount, a number of groups having the same amount, and a number of items within each group. The difference between multiplication and division is that the starting point in a division problem is the total amount. This then gets divided into equal amounts.

For example, the equation is $48 \div 8 = 6$.

48 is the total number of items, which is being divided into 8 different groups. In order to do so, 6 items go into each group. Also, 8 and 6 are interchangeable. So, the 48 items could be divided into 6 groups of 8 items each. Therefore, different types of word problems can arise from this equation.

For example, here are three types of problems:

- A boy needs 48 pieces of chalk. If there are 8 pieces in each box, how many boxes should he buy?
- A boy has 48 pieces of chalk. If each box has 6 pieces in it, how many boxes did he buy?
- A boy has partitioned all of his chalk into 8 piles, with 6 pieces in each pile. How many pieces does he have in total?

Each one of these questions involves the same equation. The third question can easily utilize the multiplication equation $8 \times 6 = ?$ instead of division. The answers are 6, 8, and 48.

## Ratios and Proportions

**Ratios** are used to show the relationship between two quantities. The ratio of oranges to apples in the grocery store may be 3 to 2. That means that for every 3 oranges, there are 2 apples. This comparison can be expanded to represent the actual number of oranges and apples. Another example may be the number of boys to girls in a math class. If the ration of boys to girls is given as 2 to 5, that means there are 2 boys to every 5 girls in the class. Ratios can also be compared if the units in each ratio are the same. The ratio of boys to girls in the math class can be compared to the ratio of boys to girls in a science class by stating which ratio is higher and which is lower.

Rates are used to compare two quantities with different units. **Unit rates** are the simplest form of rate. With unit rates, the denominator in the comparison of two units is one. For example, if someone can type at a rate of 1,000 words in 5 minutes, then his or her unit rate for typing is $\frac{1000}{5} = 200$ words in one minute or 200 words per minute. Any rate can be converted into a unit rate by dividing to make the denominator one. 1,000 words in 5 minutes has been converted into the unit rate of 200 words per minute.

Ratios and rates can be used together to convert rates into different units. For example, if someone is driving 50 kilometers per hour, that rate can be converted into miles per hour by using a ratio known as the **conversion factor**. Since the given value contains kilometers and the final answer needs to be in miles, the ratio relating miles to kilometers needs to be used. There are 0.62 miles in 1 kilometer. This, written as a ratio and in fraction form, is

$$\frac{0.62\ miles}{1\ km}$$

To convert 50km/hour into miles per hour, the following conversion needs to be set up:

$$\frac{50\ km}{hour} \times \frac{0.62\ miles}{1\ km} = 31\ miles\ per\ hour$$

The ratio between two similar geometric figures is called the **scale factor**. For example, a problem may depict two similar triangles, A and B. The scale factor from the smaller triangle A to the larger triangle B is given as 2 because the length of the corresponding side of the larger triangle, 16, is twice the corresponding side on the smaller triangle, 8. This scale factor can also be used to find the value of a missing side, $x$, in triangle A. Since the scale factor from the smaller triangle (A) to larger one (B) is 2, the larger corresponding side in triangle B (given as 25), can be divided by 2 to find the missing side in A ($x$ = 12.5). The scale factor can also be represented in the equation $2A = B$ because two times the lengths of A gives the corresponding lengths of B. This is the idea behind similar triangles.

Much like a scale factor can be written using an equation like $2A = B$, a relationship is represented by the equation $Y = kX$. X and Y are **proportional** because as values of X increase, the values of Y also increase. A relationship that is inversely proportional can be represented by the equation $Y = \frac{k}{X}$, where the value of Y decreases as the value of $x$ increases and vice versa.

Proportional reasoning can be used to solve problems involving ratios, percentages, and averages. Ratios can be used in setting up proportions and solving them to find unknowns. For example, if a student completes an average of 10 pages of math homework in 3 nights, how long would it take the student to complete 22 pages? Both ratios can be written as fractions. The second ratio would contain the unknown.

The following proportion represents this problem, where x is the unknown number of nights:

$$\frac{10\ pages}{3\ nights} = \frac{22\ pages}{x\ nights}$$

Solving this proportion entails cross-multiplying and results in the following equation: $10x = 22 \times 3$. Simplifying and solving for $x$ results in the exact solution: $x = 6.6\ nights$. The result would be rounded up to 7 because the homework would actually be completed on the 7th night.

The following problem uses ratios involving percentages:

> If 20% of the class is girls and 30 students are in the class, how many girls are in the class?

To set up this problem, it is helpful to use the common proportion:

$$\frac{\%}{100} = \frac{is}{of}$$

Within the proportion, % is the percentage of girls, 100 is the total percentage of the class, *is* is the number of girls, and *of* is the total number of students in the class. Most percentage problems can be written using this language. To solve this problem, the proportion should be set up as $\frac{20}{100} = \frac{x}{30}$, and then solved for x. Cross-multiplying results in the equation $20 * 30 = 100x$, which results in the solution $x = 6$. There are 6 girls in the class.

Problems involving volume, length, and other units can also be solved using ratios. A problem may ask for the volume of a cone to be found that has a radius, $r = 7m$ and a height, $h = 16m$. Referring to the formulas provided on the test, the volume of a cone is given as:

$$V = \pi r^2 \frac{h}{3}$$

r is the radius, and h is the height. Plugging $r = 7$ and $h = 16$ into the formula, the following is obtained:

$$V = \pi(7^2)\frac{16}{3}$$

Therefore, volume of the cone is found to be approximately 821$m^3$. Sometimes, answers in different units are sought. If this problem wanted the answer in liters, 821$m^3$ would need to be converted.

Using the equivalence statement 1$m^3$ = 1000L, the following ratio would be used to solve for liters:

$$821\text{m}^3 * \frac{1000L}{1m^3}$$

Cubic meters in the numerator and denominator cancel each other out, and the answer is converted to 821,000 liters, or $8.21 * 10^5$ L.

Other conversions can also be made between different given and final units. If the temperature in a pool is 30°C, what is the temperature of the pool in degrees Fahrenheit? To convert these units, an equation is used relating Celsius to Fahrenheit.

The following equation is used:

$$T_{°F} = 1.8T_{°C} + 32$$

Plugging in the given temperature and solving the equation for T yields the result:

$$T_{°F} = 1.8(30) + 32 = 86°F$$

Both units in the metric system and U.S. customary system are widely used.

## Complex Numbers

Some types of equations can be solved to find real answers, but this is not the case for all equations. For example, $x^2 = k$ can be solved when *k* is non-negative, but it has no real solutions when *k* is negative. Equations do have solutions if complex numbers are allowed.

**Complex numbers** are defined in the following manner: every complex number can be written as $a + bi$, where $i^2 = -1$. Thus, the solutions to the equation $x^2 = -1$ are $\pm i$.

Some examples of complex numbers include $6 + 2i$, $5 - 7i$, and $-3 + 12i$. Adding and subtracting complex numbers is similar to collecting like terms. The real numbers are added together, and the imaginary numbers are added together. For example, if the problem asks to simplify the expression $6 + 2i - 3 + 7i$, the 6 and -3 are combined to make 3, and the $2i$ and $7i$ combine to make $9i$. Multiplying and dividing complex numbers is similar to working with exponents. One rule to remember when multiplying is that $i \times i = -1$. For example, if a problem asks to simplify the expression $4i(3 + 7i)$, the $4i$ should be distributed throughout the 3 and the $7i$. This leaves the final expression $12i - 28$. The 28 is negative because $i \times i$ results in a negative number.

Complex numbers may result from solving polynomial equations using the quadratic equation. Since complex numbers result from taking the square root of a negative number, the number found under the radical in the quadratic formula—called the **determinant**—tells whether or not the answer will be real or complex. If the determinant is negative, the roots are complex. Even though the coefficients of the polynomial may be real numbers, the roots are complex.

## Counting

Counting problems involve determining how many ways one can select or arrange a specific number of objects. The basic counting principle states that if there are *m* ways of doing something and *n* ways of doing something else, then there are *m* x *n* ways of doing both. For instance, if you have 3 choices of bread and 4 choices of meat, there are 3 x 4 = 12 possible sandwiches that can be made. Another counting idea involves the topic of permutations. A **permutation** is an arrangement of a specific number of objects where repetition is not allowed, and order is important. The total number of permutations of *k* objects out of a set of *n* objects is ${}_nP_k = \frac{n!}{(n-k)!}$, where $n! = n(n-1)(n-2)\cdots 2 \times 1$. For instance, the total number of ways that 3 cars out of 9 total cars can be selected and ordered for a parade is:

$${}_9P_3 = \frac{9!}{(9-3)!} = 504$$

Notice that in this case, order does matter. If order did not matter, this situation would involve combinations. The total number of combinations of *k* objects out of a set of *n* objects is:

$${}_nC_k = \frac{n!}{(n-k)!\,k!}$$

Therefore, if order does not matter in the car example, the number of possible car selections is:

$${}_9C_3 = \frac{9!}{(9-3)!\,3!} = 84$$

## Elementary Number Theory

### Classifying Real Numbers

The mathematical number system is made up of two general types of numbers: real and complex. **Real numbers** are those that are used in normal settings, while **complex numbers** are those composed of both a real number and an imaginary one. Imaginary numbers are the result of taking the square root of -1, and $\sqrt{-1} = i$.

The real number system is often explained using a Venn diagram similar to the one below. After a number has been labeled as a real number, further classification occurs when considering the other groups in this diagram. If a number is a never-ending, non-repeating decimal, it falls in the **irrational** category. Otherwise, it is **rational**. Furthermore, if a number does not have a fractional part, it is classified as an **integer**, such as -2, 75, or 0. **Whole numbers** are an even smaller group that only includes positive integers and 0. The last group, **natural numbers**, is made up of only positive integers, such as 2, 56, or 12.

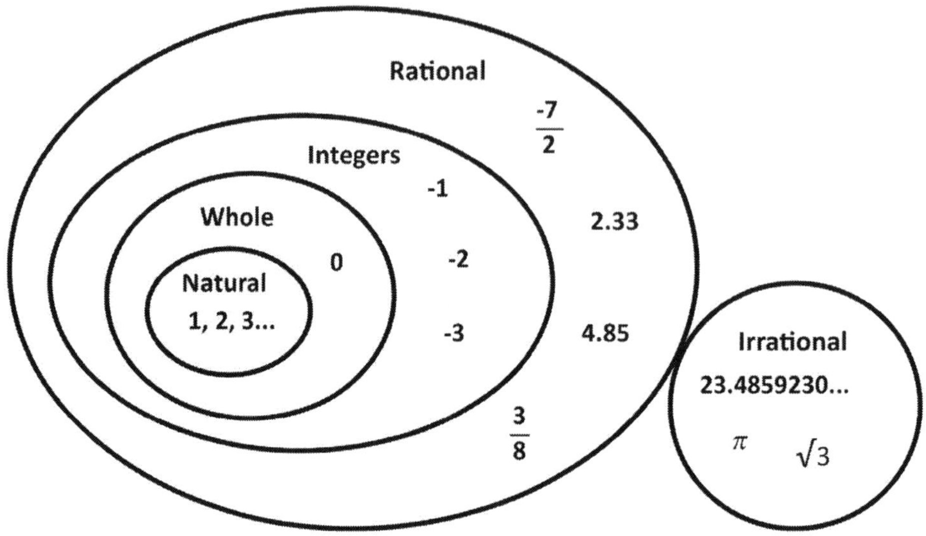


**Integers** are the whole numbers together with their negatives. They include numbers like 5, -24, 0, -6, and 15. They do not include fractions or numbers that have digits after the decimal point.

**Rational** numbers are all numbers that can be written as a fraction using integers. They are inclusive of integers, fractions, and decimals that terminate, or end (1.7, 0.04213) or repeat ($0.136\overline{5}$). A **fraction** is written as $\frac{x}{y}$ and represents the quotient of $x$ being divided by $y$. More practically, it means dividing the whole into $y$ equal parts, then taking $x$ of those parts.

Examples of rational numbers include $\frac{1}{2}$ and $\frac{5}{4}$. The number on the top is called the **numerator,** and the number on the bottom is called the **denominator**. Because every integer can be written as a fraction with a denominator of 1, (e.g. $\frac{3}{1} = 3$), every integer is also a rational number.

A **number line** typically consists of integers (...3, 2, 1, 0, -1, -2, -3...), and is used to visually represent the value of a rational number. Each rational number has a distinct position on the line determined by comparing its value with the displayed values on the line. For example, if plotting -1.5 on the number line below, it is necessary to recognize that the value of -1.5 is .5 less than -1 and .5 greater than -2. Therefore, -1.5 is plotted halfway between -1 and -2.

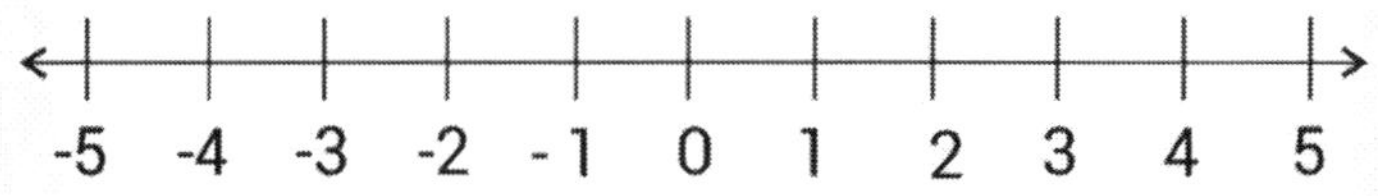


The number system that is used consists of only ten different digits or characters. However, this system is used to represent an infinite number of values. As mentioned, the **place value system** makes this infinite number of values possible. The position in which a digit is written corresponds to a given value. Starting from the decimal point (which is implied, if not physically present), each subsequent place value to the left represents a value greater than the one before it. Conversely, starting from the decimal point, each subsequent place value to the right represents a value less than the one before it.

In accordance with the **base-10 system**, the value of a digit increases by a factor of ten each place it moves to the left. For example, consider the number 7. Moving the digit one place to the left (70), increases its value by a factor of 10 ($7 \times 10 = 70$). Moving the digit two places to the left (700) increases its value by a factor of 10 twice ($7 \times 10 \times 10 = 700$). Moving the digit three places to the left (7,000) increases its value by a factor of 10 three times ($7 \times 10 \times 10 \times 10 = 7{,}000$), and so on.

Conversely, the value of a digit decreases by a factor of ten each place it moves to the right. (Note that multiplying by $\frac{1}{10}$ is equivalent to dividing by 10). For example, consider the number 40. Moving the digit one place to the right (4) decreases its value by a factor of 10 ($40 \div 10 = 4$). Moving the digit two places to the right (0.4), decreases its value by a factor of 10 twice ($40 \div 10 \div 10 = 0.4$) or ($40 \times \frac{1}{10} \times \frac{1}{10} = 0.4$). Moving the digit three places to the right (0.04) decreases its value by a factor of 10 three times ($40 \div 10 \div 10 \div 10 = 0.04$) or ($40 \times \frac{1}{10} \times \frac{1}{10} \times \frac{1}{10} = 0.04$), and so on.

## Prime and Composite Numbers

Whole numbers are classified as either prime or composite. A **prime** number can only be divided evenly by itself and one. For example, the number 11 can only be divided evenly by 11 and one; therefore, 11 is a prime number. A helpful way to visualize a prime number is to use concrete objects and try to divide them into equal piles. If dividing 11 coins, the only way to divide them into equal piles is to create 1 pile of 11 coins or to create 11 piles of 1 coin each. Other examples of prime numbers include 2, 3, 5, 7, 13, 17, and 19.

A **composite** number is any whole number that is not a prime number. A composite number is a number that can be divided evenly by one or more numbers other than itself and one. For example, the number 6 can be divided evenly by 2 and 3. Therefore, 6 is a composite number. If dividing 6 coins into equal piles, the possibilities are 1 pile of 6 coins, 2 piles of 3 coins, 3 piles of 2 coins, or 6 piles of 1 coin. Other examples of composite numbers include 4, 8, 9, 10, 12, 14, 15, 16, 18, and 20.

To determine if a number is a prime or composite number, the number is divided by every whole number greater than one and less than its own value. If it divides evenly by any of these numbers, then the number is composite. If it does not divide evenly by any of these numbers, then the number is prime. For example, when attempting to divide the number 5 by 2, 3, and 4, none of these numbers divide evenly. Therefore, 5 must be a prime number.

## Odd and Even Numbers

**Even numbers** are all divisible by the number 2. **Odd numbers** are not divisible by 2, and an odd quantity of items cannot be paired up into groups of 2 without having 1 item leftover. Examples of even numbers are 2, 4, 6, 20, 100, 242, etc. Examples of odd numbers are 1, 3, 5, 27, 99, 333, etc.

## Factors, Multiples, and Divisibility

The **Fundamental Theorem of Arithmetic** states that any integer greater than 1 is either a prime number or can be written as a unique product of prime numbers. Factors can be used to find the combination of numbers to multiply to produce an integer that is not prime. The factors of a number are all integers that can be multiplied by another integer to produce the given number. For example, 2 is multiplied by 3 to produce 6. Therefore, 2 and 3 are both factors of 6. Similarly, $1 \times 6 = 6$ and $2 \times 3 = 6$, so 1, 2, 3, and 6 are all factors of 6. Another way to explain a factor is to say that a given number divides evenly by each of its factors to produce an integer. For example, 6 does not divide evenly by 5. Therefore, 5 is not a factor of 6.

Essentially, **factors** are the numbers multiplied to achieve a product. Thus, every product in a multiplication equation has, at minimum, two factors. Of course, some products will have more than two factors. For the sake of most discussions, assume that factors are positive integers.

To find a number's factors, start with 1 and the number itself. Then divide the number by 2, 3, 4, and so on, seeing if any divisors can divide the number without a remainder, keeping a list of those that do. Stop upon reaching either the number itself or another factor.

Let's find the factors of 45. Start with 1 and 45. Then try to divide 45 by 2, which fails. Now divide 45 by 3. The answer is 15, so 3 and 15 are now factors. Dividing by 4 doesn't work, and dividing by 5 leaves 9. Lastly, dividing 45 by 6, 7, and 8 all don't work. The next integer to try is 9, but this is already known to be a factor, so the factorization is complete. The factors of 45 are 1, 3, 5, 9, 15 and 45.

A **common factor** is a factor shared by two numbers. Let's take 45 and 30 and find the common factors:

The factors of 45 are: 1, 3, 5, 9, 15, and 45.
The factors of 30 are: 1, 2, 3, 5, 6, 10, 15, and 30.
The common factors are 1, 3, 5, and 15.

The **greatest common factor** is the largest number among the shared, common factors. From the factors of 45 and 30, the common factors are 3, 5, and 15. Thus, 15 is the greatest common factor, as it's the largest number.

**Multiples** of a given number are found by taking that number and multiplying it by any other whole number. For example, 3 is a factor of 6, 9, and 12. Therefore, 6, 9, and 12 are multiples of 3. The multiples of any number are an infinite list. For example, the multiples of 5 are 5, 10, 15, 20, and so on. This list continues without end. A list of multiples is used in finding the **least common multiple**, or **LCM,** for fractions when a common denominator is needed. The denominators are written down and their multiples listed until a common number is found in both lists. This common number is the LCM.

If two numbers share no factors besides 1 in common, then their least common multiple will be simply their product. If two numbers have common factors, then their least common multiple will be their product divided by their greatest common factor. This can be visualized by the formula $LCM = \frac{x \times y}{GCF}$, where *x* and *y* are some integers and LCM and GCF are their least common multiple and greatest common factor, respectively.

**Prime factorization** breaks down each factor of a whole number until only prime numbers remain. All composite numbers can be factored into prime numbers. For example, the prime factors of 12 are 2, 2, and 3 ($2 \times 2 \times 3 = 12$). To produce the prime factors of a number, the number is factored, and any composite numbers are continuously factored until the result is the product of prime factors only. A factor tree, such as the one below, is helpful when exploring this concept.

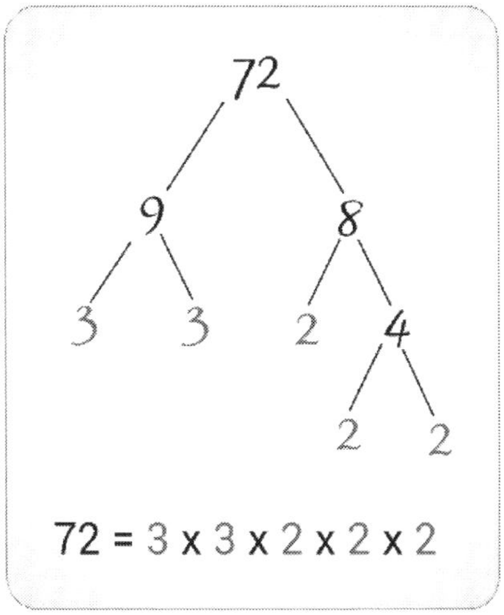


Let's break 129 down into its prime factors. First, the factors are 3 and 43. Both 3 and 43 are prime numbers, so we're done. But if 43 was not a prime number, then it would also need to be factorized until all of the factors are expressed as prime numbers.

## Matrices

**Matrices** can be used to represent linear equations, solve systems of equations, and manipulate data to simulate change. Matrices consist of numerical entries in both rows and columns. The following matrix A is a $3\ x\ 4$ matrix because it has three rows and four columns:

$$A = \begin{bmatrix} 3 & 2 & -5 & 3 \\ 3 & 6 & 2 & -5 \\ -1 & 3 & 7 & 0 \end{bmatrix}$$

Matrices can be added or subtracted only if they have the same dimensions. For example, the following matrices can be added by adding corresponding matrix entries:

$$\begin{bmatrix} 3 & 4 \\ 2 & -6 \end{bmatrix} + \begin{bmatrix} -1 & 4 \\ 4 & 2 \end{bmatrix} = \begin{bmatrix} 2 & 8 \\ 6 & -4 \end{bmatrix}$$

Multiplication can also be used to manipulate matrices. **Scalar multiplication** involves multiplying a matrix by a constant. Each matrix entry needs to be multiplied times the constant. The following example shows a $3 \times 2$ matrix being multiplied by the constant 6:

$$6 \times \begin{bmatrix} 3 & 4 \\ 2 & -6 \\ 1 & 0 \end{bmatrix} = \begin{bmatrix} 18 & 24 \\ 12 & -36 \\ 6 & 0 \end{bmatrix}$$

**Matrix multiplication** of two matrices involves finding multiple dot products. The **dot product** of a row and column is the sum of the products of each corresponding row and column entry. In the following example, a $2 \times 2$ matrix is multiplied by a $2 \times 2$ matrix.

The dot product of the first row and column is:

$$(1 \times 2) + (2 \times 1) = (2) + (2) = 4$$

$$\begin{bmatrix} 1 & 2 \\ 3 & 4 \end{bmatrix} \text{X} \begin{bmatrix} 2 & 0 \\ 1 & 2 \end{bmatrix} = \begin{bmatrix} 4 & 4 \\ 10 & 8 \end{bmatrix}$$

The same process is followed to find the other three values in the solution matrix. Matrices can only be multiplied if the number of columns in the first matrix equals the number of rows in the second matrix. The previous example is also an example of square matrix multiplication because they are both square matrices. A **square matrix** has the same number of rows and columns. For square matrices, the order in which they are multiplied does matter. Therefore, matrix multiplication does not satisfy the commutative property. It does, however, satisfy the associative and distributive properties.

Another transformation of matrices can be found by using the **identity matrix**—also referred to as the "I" matrix. The identity matrix is similar to the number one in normal multiplication. The identity matrix is a square matrix with ones in the diagonal spots and zeros everywhere else. The identity matrix is also the result of multiplying a matrix by its inverse. This process is similar to multiplying a number by its reciprocal.

The **zero matrix** is also a matrix acting as an additive identity. The zero matrix consists of zeros in every entry. It does not change the values of a matrix when using addition.

Given a system of linear equations, a matrix can be used to represent the entire system. Operations can then be performed on the matrix to solve the system. The following system offers an example:

$$\begin{aligned} x + y + z &= 6 \\ y + 5z &= -4 \\ 2x + 5y - z &= 27 \end{aligned}$$

There are three variables and three equations. The coefficients in the equations can be used to form a 3 x 3 matrix:

$$\begin{bmatrix} 1 & 1 & 1 \\ 0 & 1 & 5 \\ 2 & 5 & -1 \end{bmatrix}$$

The number of rows equals the number of equations, and the number of columns equals the number of variables. The numbers on the right side of the equations can be turned into a $3 \times 1$ matrix. That matrix is shown here:

$$\begin{bmatrix} 6 \\ -4 \\ 27 \end{bmatrix}$$

It can also be referred to as a **vector**. The variables are represented in a matrix of their own:

$$\begin{bmatrix} x \\ y \\ z \end{bmatrix}$$

The system can be represented by the following matrix equation:

$$\begin{bmatrix} 1 & 1 & 1 \\ 0 & 2 & 5 \\ 2 & 5 & -1 \end{bmatrix}\begin{bmatrix} x \\ y \\ z \end{bmatrix}=\begin{bmatrix} 6 \\ -4 \\ 27 \end{bmatrix}$$

Simply, this is written as $AX = B$. By using the inverse of a matrix, the solution can be found: $X = A^{-1}B$. Once the inverse of A is found using operations, it is then multiplied by B to find the solution to the system: $x = 5, y = 3,\text{and } z = -2$.

The determinant of a 2 x 2 matrix is the following:

$$|A| = \begin{vmatrix} a & b \\ c & d \end{vmatrix} = ad - bc$$

It is a number related to the size of the matrix. The absolute value of the determinant of matrix A is equal to the area of a parallelogram with vertices (0, 0), (a, b), (c, d), and (a+b, c+d).

## Sequences and Series

A **sequence** is an enumerated set of numbers, and each term or member is defined by the number that exists within the sequence. It can have either a finite or infinite number of terms, and a sequence is written as $\{a_n\}$, where $a_n$ is the nth term of the sequence. An example of an infinite sequence is:

$$\left\{\frac{n+1}{n^2}\right\}_{n=1}^{\infty}$$

Its first three terms are found by evaluating at n=1, 2, and 3 to get 2, $\frac{3}{4}$, and $\frac{4}{9}$. Limits of infinite sequences, if they exist, can be found in a similar manner as finding infinite limits of functions. $n$ needs to be treated as a variable, and then $\lim_{n\to\infty}\frac{n+1}{n^2}$ can be evaluated, resulting in 0.

An **infinite series** is the sum of an infinite sequence. For example, $\sum_{n=1}^{\infty}\frac{n+1}{n^2}$ is the infinite series of the sequence given above. Partial sums are sums of a finite number of terms. For example, $s_{10}$ represents the sum of the first 10 terms, and in general, $s_n$ represents the sum of the first n terms. An infinite series can either converge or diverge. If the sum of an infinite series is a finite number, the series is said to **converge**. Otherwise, it **diverges**. In the general infinite series $\sum a_n$, If $\lim_{n\to\infty} a_n \neq 0$ or does not exist, the series diverges. However, if $\lim_{n\to\infty} a_n = 0$, the series does not necessarily converge.

Several tests exist that determine whether a series converges:

- The Absolute Convergence Test states that if $\sum |a_n|$ converges, then $\sum a_n$ converges.
- The Integral Test states that if $f(n) = a_n$ is a positive, continuous, decreasing function, then $\sum a_n$ is convergent if and only if $\int_1^\infty f(x)dx$ is convergent. The geometric series $\sum ar^{n-1}$ is convergent if $|r| < 1$ and its sum is equal to $\frac{a}{1-r}$. If $|r| \geq 1$, the geometric series is divergent.
- The Limit Comparison Tests compares two infinite series $\sum a_n$ and $\sum b_n$ with positive terms. If $\sum b_n$converges and $a_n \leq b_n$ for all n, then $\sum a_n$ converges. If $\sum b_n$diverges and $a_n \geq b_n$for all n, then $\sum a_n$ diverges. If $\lim_{n\to\infty} \frac{a_n}{b_n} = c$, where c is a finite, positive number, then either both series converge or diverge.
- The Alternating Series Test states that if $b_{n+1} \geq b_n$ for all n and $\lim_{n\to\infty} b_n = 0$, then the series:

  $$\sum(-1)^{n-1}b_n$$

- The Ratio Test states that if the limit of the ratio of consecutive terms $a_{n+1}/a_n$ is less than 1, then the series is convergent. If the ratio is greater than 1, the series is divergent. If the limit is equal to 1, the test is inconclusive.
- The Root Test states that if $\lim_{n\to\infty} \sqrt[n]{|a_n|} < 1$, the series converges. If the same limit is greater than 1, the series diverges, and if the limit equals 1, the test is inconclusive.

## Interest

One exponential formula that is commonly used is $A = Pe^{rt}$. This is the formula for **continually compounded interest**. $A$ is the value of the investment after the time, $t$, in years. $P$ is the initial amount of the investment, $r$ is the interest rate, and $e$ is the constant equal to approximately 2.718. Given an initial amount of $200 and a time of 3 years, if interest is compounded continuously at a rate of 6%, the total investment value can be found by plugging each value into the formula. The invested value at the end is $239.44. In more complex problems, the final investment may be given, and the rate may be the unknown. In this case, the formula becomes $239.44 = 200e^{r3}$. Solving for $r$ requires isolating the exponential expression on one side by dividing by 200, yielding the equation $1.20 = e^{r3}$. Taking the natural log of both sides results in $\ln(1.2) = r3$. Using a calculator to evaluate the logarithmic expression, $r = 0.06 = 6\%$.

When considering the economic value of an investment and neglecting inflation, the future value, F, of a present value of P dollars, invested for n years at a fixed interest rate, r%, is given by the **single payment compound amount factor** as:

$F = P(1 + {}^{r}/_{100})^n$, which is often solved in tables titled "(F/P, i%, n)."

Inversely, the **single payment present worth factor (SPPWF)** states that the present value, P, of a future value of F dollars, n years from now at the same rates is:

$P = F(1 + {}^{r}/_{100})^{-n}$, often solved in "(P/F, i%, n)" tables.

Any future costs over any amount of time can be expressed in current dollars by applying the SPPWF equation as shown in the following example, worked out in the table below.

Consider seven equal annual payments of $5 million each, and determine the net present value of the payments with interest rates of 4 and 8 percent:

$$P = F(1 + {}^{r}/_{100})^{-n}$$

| Year, n | Future Cost | Present Value: 4% Interest | Present Value: 8% Interest |
|---|---|---|---|
| 0 | $5,000,000 | $5,000,000 | $5,000,000 |
| 1 | $5,000,000 | $4,807,692 | $4,629,630 |
| 2 | $5,000,000 | $4,622,781 | $4,286,694 |
| 3 | $5,000,000 | $4,444,982 | $3,969,161 |
| 4 | $5,000,000 | $4,274,021 | $3,675,149 |
| 5 | $5,000,000 | $4,109,636 | $3,402,916 |
| 6 | $5,000,000 | $3,951,573 | $3,150,848 |
| 7 | $5,000,000 | $3,799,589 | $2,917,452 |
| **Total** | **$40,000,000** | **$35,010,274** | **$31,031,850** |

## Patterns

Patterns within a sequence can come in 2 distinct forms: the items (shapes, numbers, etc.) either repeat in a constant order, or the items change from one step to another in some consistent way. The **core** is the smallest unit, or number of items, that repeats in a repeating pattern. For example, the pattern oo▲oo▲o… has a core that is oo▲. Knowing only the core, the pattern can be extended. Knowing the number of steps in the core allows the identification of an item in each step without drawing/writing the entire pattern out. For example, suppose the tenth item in the previous pattern must be determined. Because the core consists of three items (oo▲), the core repeats in multiples of 3. In other words, steps 3, 6, 9, 12, etc. will be ▲ completing the core with the core starting over on the next step. For the above example, the 9th step will be ▲ and the 10th will be o.

The most common patterns in which each item changes from one step to the next are arithmetic and geometric sequences. An **arithmetic sequence** is one in which the items increase or decrease by a constant difference. In other words, the same thing is added or subtracted to each item or step to produce the next. To determine if a sequence is arithmetic, determine what must be added or subtracted to step one to produce step two. Then, check if the same thing is added/subtracted to step two to produce step three. The same thing must be added/subtracted to step three to produce step four, and so on. Consider the pattern 13, 10, 7, 4 . . . To get from step one (13) to step two (10) by adding or subtracting requires subtracting by 3. The next step is checking if subtracting 3 from step two (10) will produce step three (7), and subtracting 3 from step three (7) will produce step four (4). In this case, the pattern holds true. Therefore, this is an arithmetic sequence in which each step is produced by subtracting 3 from the previous step. To extend the sequence, 3 is subtracted from the last step to produce the next. The next three numbers in the sequence are 1, -2, -5.

A **geometric sequence** is one in which each step is produced by multiplying or dividing the previous step by the same number. To determine if a sequence is geometric, decide what step one must be multiplied or divided by to produce step two. Then check if multiplying or dividing step two by the same number produces step three, and so on. Consider the pattern 2, 8, 32, 128 . . . To get from step one (2) to step two (8) requires multiplication by 4. The next step determines if multiplying step two (8) by 4 produces step three (32), and multiplying step three (32) by 4 produces step four (128). In this case, the pattern holds true. Therefore, this is a geometric sequence in which each step is produced by multiplying the previous step by 4. To extend the sequence, the last step is multiplied by 4 and repeated. The next three numbers in the sequence are 512; 2,048; 8,192.

Although arithmetic and geometric sequences typically use numbers, these sequences can also be represented by shapes. For example, an arithmetic sequence could consist of shapes with three sides, four sides, and five sides (add one side to the previous step to produce the next). A geometric sequence could consist of eight blocks, four blocks, and two blocks (each step is produced by dividing the number of blocks in the previous step by 2).

## *Conjectures, Predictions, or Generalizations Based on Patterns*

An arithmetic or geometric sequence can be written as a formula and used to determine unknown steps without writing out the entire sequence. (Note that a similar process for repeating patterns is covered in the previous section.) An arithmetic sequence progresses by a **common difference**. To determine the common difference, any step is subtracted by the step that precedes it. In the sequence 4, 9, 14, 19 . . . the common difference, or *d*, is 5. By expressing each step as $a_1$, $a_2$, $a_3$, etc., a formula can be written to represent the sequence. $a_1$ is the first step. To produce step two, step 1 ($a_1$) is added to the common difference (*d*):

$$a_2 = a_1 + d$$

To produce step three, the common difference (*d*) is added twice to $a_1$:

$$a_3 = a_1 + 2d$$

To produce step four, the common difference (*d*) is added three times to $a_1$:

$$a_4 = a_1 + 3d$$

Following this pattern allows a general rule for arithmetic sequences to be written. For any term of the sequence ($a_n$), the first step ($a_1$) is added to the product of the common difference (*d*) and one less than the step of the term ($n - 1$):

$$a_n = a_1 + (n - 1)d$$

Suppose the 8th term ($a_8$) is to be found in the previous sequence. By knowing the first step ($a_1$) is 4 and the common difference (*d*) is 5, the formula can be used:

$$a_n = a_1 + (n-1)d$$

$$a_8 = 4 + (7)5$$

$$a_8 = 39$$

In a geometric sequence, each step is produced by multiplying or dividing the previous step by the same number. The **common ratio**, or (*r*), can be determined by dividing any step by the previous step. In the sequence 1, 3, 9, 27 . . . the common ratio (*r*) is 3($\frac{3}{1} = 3$ or $\frac{9}{3} = 3$ or $\frac{27}{9} = 3$). Each successive step can be expressed as a product of the first step ($a_1$) and the common ratio (*r*) to some power. For example:

$$a_2 = a_1 \times r$$

$$a_3 = a_1 \times r \times r$$

$$a_3 = a_1 \times r^2$$

$$a_4 = a_1 \times r \times r \times r$$

or

$$a_4 = a\_1 \times r^3$$

Following this pattern, a general rule for geometric sequences can be written. For any term of the sequence ($a_n$), the first step ($a_1$) is multiplied by the common ratio (*r*) raised to the power one less than the step of the term ($n-1$):

$$a_n = a_1 \times r^{(n-1)}$$

Suppose for the previous sequence, the 7th term ($a_7$) is to be found. Knowing the first step ($a_1$) is one, and the common ratio (*r*) is 3, the formula can be used:

$$a_n = a_1 \times r^{(n-1)}$$

$$a_7 = (1) \times 3^6 \rightarrow a_7 = 729$$

### *Corresponding Terms of Two Numerical Patterns*

When given two numerical patterns, the corresponding terms should be examined to determine if a relationship exists between them. Corresponding terms between patterns are the pairs of numbers that appear in the same step of the two sequences. Consider the following patterns 1, 2, 3, 4 . . . and 3, 6, 9, 12 . . . The corresponding terms are: 1 and 3; 2 and 6; 3 and 9; and 4 and 12. To identify the relationship, each pair of corresponding terms is examined and the possibilities of performing an operation (+, −, ×, ÷) to the term from the first sequence to produce the corresponding term in the second sequence are determined. In this case:

$1 + 2 = 3$ or $1 \times 3 = 3$

$2 + 4 = 6$ or $2 \times 3 = 6$

$3 + 6 = 9$ or $3 \times 3 = 9$

$4 + 8 = 12$ or $4 \times 3 = 12$

The consistent pattern is that the number from the first sequence multiplied by 3 equals its corresponding term in the second sequence. By assigning each sequence a label (input and output) or variable (*x* and *y*), the relationship can be written as an equation. If the first sequence represents the inputs, or *x*, and the second sequence represents the outputs, or *y*, the relationship can be expressed as: $y = 3x$.

Consider the following sets of numbers:

| | | | | |
|---|---|---|---|---|
| **a** | 2 | 4 | 6 | 8 |
| **b** | 6 | 8 | 10 | 12 |

To write a rule for the relationship between the values for *a* and the values for *b*, the corresponding terms (2 and 6; 4 and 8; 6 and 10; 8 and 12) are examined. The possibilities for producing *b* from *a* are:

$2 + 4 = 6$ or $2 \times 3 = 6$

$4 + 4 = 8$ or $4 \times 2 = 8$

$6 + 4 = 10$

$8 + 4 = 12$ or $8 \times 1.5 = 12$

The consistent pattern is that adding 4 to the value of *a* produces the value of *b*. The relationship can be written as the equation $a + 4 = b$.

## Vectors

A vector can be thought of as a list of numbers. These can be thought of as an abstract list of numbers, or else as giving a location in a space. For example, the coordinates $(x, y)$ for points in the Cartesian plane are vectors. Each entry in a vector can be referred to by its location in the list: first, second, and so on. The total length of the list is the **dimension** of the vector. A vector is often denoted as such by putting an arrow on top of it, e.g. $\vec{v} = (v_1, v_2, v_3)$

### Adding Vectors

There are two basic operations for vectors. First, two vectors can be added together. Let:

$$\vec{v} = (v_1, v_2, v_3), \vec{w} = (w_1, w_2, w_3)$$

The the sum of the two vectors is defined to be:

$$\vec{v} + \vec{w} = (v_1 + w_1, v_2 + w_2, v_3 + w_3)$$

Subtraction of vectors can be defined similarly.

Vector addition can be visualized in the following manner. First, visualize each vector as an arrow. Then place the base of one arrow at the tip of the other arrow. The tip of this first arrow now hits some point in the space, and there will be an arrow from the origin to this point. This new arrow corresponds to the new vector. In subtraction, we reverse the direction of the arrow being subtracted.

For example, consider adding together the vectors (-2, 3) and (4, 1). The new vector will be (-2+4, 3+1), or (2, 4). Graphically, this may be pictured in the following manner.

## Performing Scalar Multiplication

The second basic operation for vectors is called **scalar multiplication**. Scalar multiplication allows us to multiply any vector by any real number, which is denoted here as a scalar. Let $\vec{v} = (v_1, v_2, v_3)$, and let *a* be an arbitrary real number. Then the scalar multiple $a\vec{v} = (av_1, av_2, av_3)$. Graphically, this corresponds to changing the length of the arrow corresponding to the vector by a factor, or scale, of *a*. That is why the real number is called a scalar in this instance.

As an example, let $\vec{v} = \left(2, -1, \frac{1}{3}\right)$

Then:

$$3\vec{v} = \left(3 \cdot 2, 3(-1), 3 \cdot \frac{1}{3}\right) = (6, -3, 1)$$

Note that scalar multiplication is distributive over vector addition, meaning that:

$$a(\vec{v} + \vec{w}) = a\vec{v} + a\vec{w}$$

## The Basic Properties of Vectors

Two vectors are equal if, and only if, their individual components are equal. For example, $< 1,2 > = < 1,2 >$ but $< 1,2 > \neq < 2,1 >$. Also, vector addition is performed component-wise. Such an example is:

$$< 1,2 > + < 2,3 \geq < 3,5 >$$

The zero vector is defined to be the vector containing only components equal to 0, and any vector plus a zero vector equals itself. Hence, the zero vector is the additive identity. Scalar multiplication is performed by multiplying each component by the scalar. For example:

$$3 \cdot < 1,2 \geq < 3,6 >$$

Scalar multiplication and addition can be used to prove that the distributive property holds within vector addition and scalar addition. Vector multiplication is defined using the **dot product,** which is also known as the **scalar product**. The result of a dot product is a scalar. Each corresponding component is multiplied, and then the sum of all products is found. For example:

$$< 1,2 > \cdot < 2,3 \geq 1 \cdot 2 + 2 \cdot 3$$

$$2 + 6 = 8$$

Alternatively, the dot product is defined to be the product of the magnitudes of each vector and the cosine of the angle between the two vectors. Therefore, if two vectors are perpendicular, their dot product is equal to zero. Finally, two vectors are parallel if they are scalar multiples of each other.

# Practice Questions

1. What is the solution to $4 \times 7 + (25 - 21)^2 \div 2$?
   a. 512
   b. 36
   c. 60.5
   d. 22
   e. 16

2. Is the series $\sum_{k=0}^{\infty}(-1)^k\left(\frac{2}{3}\right)^k$ convergent or divergent? If convergent, find its sum.
   a. Divergent
   b. Convergent, $\frac{3}{5}$
   c. Convergent, $\frac{5}{3}$
   d. Convergent, $\frac{2}{3}$
   e. Convergent, $\frac{3}{2}$

3. Find the determinant of the matrix $\begin{bmatrix} -4 & 2 \\ 3 & -1 \end{bmatrix}$.
   a. -10
   b. -2
   c. 0
   d. 2
   e. -1

4. Five of six numbers have a sum of 25. The average of all six numbers is 6. What is the sixth number?
   a. 8
   b. 10
   c. 13
   d. 12
   e. 11

5. What is the next number in the following series: $1, 3, 6, 10, 15, 21, \ldots$?
   a. 26
   b. 27
   c. 28
   d. 29
   e. 30

6. What are all the factors of 12?
   a. 12, 24, 36
   b. 1, 2, 4, 6, 12
   c. 12, 24, 36, 48
   d. 1, 2, 3, 4, 6, 12
   e. 0, 1, 12

7. Which of the following augmented matrices represents the system of equations below?

$$2x - 3y + z = -5$$
$$4x - y - 2z = -7$$
$$-x + 2z = -1$$

a. $\begin{bmatrix} 2 & -3 & 1 & 5 \\ 4 & -1 & 0 & -7 \\ -1 & 0 & 2 & 1 \end{bmatrix}$

b. $\begin{bmatrix} 2 & 4 & -1 \\ -3 & -1 & 0 \\ 1 & -2 & 2 \\ -5 & -7 & -1 \end{bmatrix}$

c. $\begin{bmatrix} 2 & 4 & -1 & -5 \\ -3 & -1 & 0 & -7 \\ 2 & -2 & 2 & -1 \end{bmatrix}$

d. $\begin{bmatrix} 2 & -3 & 1 \\ 4 & -1 & -2 \\ -1 & 0 & 2 \end{bmatrix}$

e. $\begin{bmatrix} 2 & -3 & 1 & -5 \\ 4 & -1 & -2 & -7 \\ -1 & 0 & 2 & -1 \end{bmatrix}$

8. Which of the following correctly arranges the numbers from least to greatest value?
$0.85, \frac{4}{5}, \frac{2}{3}, \frac{91}{100}$

a. $0.85, \frac{4}{5}, \frac{2}{3}, \frac{91}{100}$

b. $\frac{4}{5}, 0.85, \frac{91}{100}, \frac{2}{3}$

c. $\frac{2}{3}, \frac{4}{5}, 0.85, \frac{91}{100}$

d. $0.85, \frac{91}{100}, \frac{4}{5}, \frac{2}{3}$

e. $\frac{4}{5}, \frac{2}{3}, 0.85, \frac{91}{100}$

9. What are the first four terms of the series $\left\{\frac{(-1)^{n+1}}{n^2+5}\right\}_{n=0}^{\infty}$?

a. $\frac{1}{6}, \frac{1}{9}, \frac{1}{14}, \frac{1}{19}$

b. $\frac{1}{6}, \frac{-1}{9}, \frac{1}{14}, \frac{-1}{19}$

c. $\frac{-1}{5}, \frac{1}{6}, \frac{-1}{9}, \frac{1}{14}$

d. $\frac{1}{5}, \frac{1}{6}, \frac{1}{9}, \frac{1}{14}$

e. $\frac{1}{5}, -\frac{1}{6}, -\frac{1}{9}, \frac{1}{14}$

10. In a school with 300 students, there are 10 students with red hair, 50 students with black hair, 180 students with brown hair, and 60 students with blonde hair. What is the ratio of blonde hair to brown hair?

a. 3:1
b. 2:1
c. 1:3
d. 1:2
e. 1:5

# Answer Explanations

**1. B:** To solve this correctly, keep in mind the order of operations with the mnemonic PEMDAS (Please Excuse My Dear Aunt Sally). This stands for Parentheses, Exponents, Multiplication, Division, Addition, Subtraction. Taking it step by step, solve inside the parentheses first:

$$4 \times 7 + (4)^2 \div 2$$

Then, apply the exponent:

$$4 \times 7 + 16 \div 2$$

Multiplication and division are both performed next:

$$28 + 8 = 36$$

Addition and subtraction are done last. The solution is 36.

**2. B:** The given series is a geometric series because it can be written as $\sum_{k=0}^{\infty}\left(\frac{-2}{3}\right)^k$, and it is convergent because $|r| = \frac{2}{3} < 1$. Its sum is:

$$\frac{1}{1-(-\frac{2}{3})} = \frac{3}{5}$$

**3. B:** The determinant of a 2 x 2 matrix is $ad - bc$. The calculation is:

$$-4(-1) - 2(3) = 4 - 6 = -2$$

**4. E:** The average is calculated by adding all six numbers, then dividing by 6. The first five numbers have a sum of 25. If the total divided by 6 is equal to 6, then the total itself must be 36. The sixth number must be 36 – 25 = 11.

**5. C:** Each subsequent number in the sequence adds one more than the difference between the previous two. For example:

$$10 - 6 = 4, 4 + 1 = 5$$

Therefore, the next number after 10 is $10 + 5 = 15$. Going forward:

$$21 - 15 = 6, 6 + 1 = 7$$

The next number is $21 + 7 = 28$. Therefore, the difference between numbers is the set of whole numbers starting at 2: 2, 3, 4, 5, 6, 7, ....

**6. D:** 1, 2, 3, 4, 6, 12. A given number divides evenly by each of its factors to produce an integer (no decimals). The number 5, 7, 8, 9, 10, 11 (and their opposites) do not divide evenly into 12. Therefore, these numbers are not factors.

**7. E:** The augmented matrix that represents the system of equations has dimensions $4 \times 3$ because there are three equations with three unknowns. The coefficients of the variables make up the first three columns, and the last column is made up of the numbers to the right of the equals sign. This system can be solved by reducing the matrix to row-echelon form, where the last column gives the solution for the unknown variables.

**8. C:** The first step is to depict each number using decimals. $\frac{91}{100}$ = 0.91

Dividing the numerator by denominator of $\frac{4}{5}$ to convert it to a decimal yields 0.80, while $\frac{2}{3}$ becomes 0.66 recurring. Rearrange each expression in ascending order, as found in answer C.

**9. C:** The numerator in the following sequence indicates that the sign of each term changes from term to term:

$$\left\{\frac{(-1)^{n+1}}{n^2+5}\right\}_{n=0}^{\infty}$$

The first term is negative because $n = 0$ and:

$$-1^{n+1} = -1^1 = -1$$

Therefore, the second term is positive. The third term is negative, etc. The denominator is evaluated like a function for plugging in the various n value. For example, the denominator of the first term, when $n = 0$, is $0^2 + 5 = 5$.

**10. C:** There are 60 students with blonde hair and 180 students with brown hair. The ratio would be set up as blonde:brown, so 60:180. When reduced this is 1:3. This means that for every 1 student with blonde hair, there are 3 with brown hair.

# Algebra and Functions

## Expressions

An **algebraic expression** is a statement about an unknown quantity expressed in mathematical symbols. A variable is used to represent the unknown quantity, usually denoted by a letter. An **equation** is a statement in which two expressions (at least one containing a variable) are equal to each other. An algebraic expression can be thought of as a mathematical phrase and an equation can be thought of as a mathematical sentence.

Algebraic expressions and equations both contain numbers, variables, and mathematical operations. The following are examples of algebraic expressions: $5x + 3, 7xy - 8(x^2 + y)$, and $\sqrt{a^2 + b^2}$. An expression can be simplified or evaluated for given values of variables. The following are examples of equations: $2x + 3 = 7$, $a^2 + b^2 = c^2$, and $2x + 5 = 3x - 2$. An equation contains two sides separated by an equal sign. **Equations** can be solved to determine the value(s) of the variable for which the statement is true.

### Parts of Expressions

Algebraic expressions consist of variables, numbers, and operations. A **term** of an expression is any combination of numbers and/or variables, and terms are separated by addition and subtraction. For example, the expression $5x^2 - 3xy + 4 - 2$ consists of 4 terms: $5x^2$, -3*xy*, 4*y*, and -2. Note that each term includes its given sign (+ or $-$). The **variable** part of a term is a letter that represents an unknown quantity. The **coefficient** of a term is the number by which the variable is multiplied. For the term 4*y*, the variable is y, and the coefficient is 4. Terms are identified by the power (or exponent) of its variable.

A number without a variable is referred to as a **constant**. If the variable is to the first power ($x^1$ or simply *x*), it is referred to as a linear term. A term with a variable to the second power ($x^2$) is quadratic, and a term to the third power ($x^3$) is cubic. Consider the expression $x^3 + 3x - 1$. The constant is -1. The linear term is 3*x*. There is no quadratic term. The cubic term is $x^3$.

An algebraic expression can also be classified by how many terms exist in the expression. Any like terms should be combined before classifying. A **monomial** is an expression consisting of only one term. Examples of monomials are: 17, 2*x*, and $-5ab^2$. A **binomial** is an expression consisting of two terms separated by addition or subtraction. Examples include $2x - 4$ and $-3y^2 + 2y$. A **trinomial** consists of 3 terms. For example, $5x^2 - 2x + 1$ is a trinomial.

### Adding and Subtracting Linear Algebraic Expressions

An algebraic expression is simplified by combining like terms. A **term** is a number, variable, or product of a number and variables separated by addition and subtraction. For the algebraic expression $3x^2 - 4x + 5 - 5x^2 + x - 3$, the terms are $3x^2$, -4*x*, 5, -$5x^2$, *x*, and -3. **Like terms** have the same variables raised to the same powers (exponents). The like terms for the previous example are $3x^2$ and -$5x^2$, -4x and x, 5 and -3. To combine like terms, the coefficients (numerical factor of the term including sign) are added and the variables and their powers are kept the same. Note that if a coefficient is not written, it is an implied coefficient of 1 ($x = 1x$). The previous example will simplify to:

$$-2x^2 - 3x + 2$$

When adding or subtracting algebraic expressions, each expression is written in parenthesis. The negative sign is distributed when necessary, and like terms are combined. Consider the following: add $2a + 5b - 2$ to $a - 2b + 8c - 4$. The sum is set as follows:

$$(a - 2b + 8c - 4) + (2a + 5b - 2)$$

In front of each set of parentheses is an implied positive one, which, when distributed, does not change any of the terms. Therefore, the parentheses are dropped and like terms are combined:

$$a - 2b + 8c - 4 + 2a + 5b - 2$$

$$3a + 3b + 8c - 6$$

Consider the following problem: Subtract $2a + 5b - 2$ from $a - 2b + 8c - 4$. The difference is set as follows:

$$(a - 2b + 8c - 4) - (2a + 5b - 2)$$

The implied one in front of the first set of parentheses will not change those four terms. However, distributing the implied -1 in front of the second set of parentheses will change the sign of each of those three terms:

$$a - 2b + 8c - 4 - 2a - 5b + 2$$

Combining like terms yields the simplified expression:

$$-a - 7b + 8c - 2$$

## Distributive Property

The **distributive property** states that multiplying a sum (or difference) by a number produces the same result as multiplying each value in the sum (or difference) by the number and adding (or subtracting) the products. Using mathematical symbols, the distributive property states:

$$a(b + c) = ab + ac$$

The expression $4(3 + 2)$ is simplified using the order of operations. Simplifying inside the parenthesis first produces $4 \times 5$, which equals 20. The expression $4(3 + 2)$ can also be simplified using the distributive property:

$$4(3 + 2) = 4 \times 3 + 4 \times 2$$

$$12 + 8 = 20$$

Consider the following example: $4(3x - 2)$. The expression cannot be simplified inside the parenthesis because 3*x* and -2 are not like terms and therefore cannot be combined. However, the expression can be simplified by using the distributive property and multiplying each term inside of the parenthesis by the term outside of the parenthesis: $12x - 8$. The resulting equivalent expression contains no like terms, so it cannot be further simplified.

Consider the expression:

$$(3x + 2y + 1) - (5x - 3) + 2(3y + 4)$$

Again, there are no like terms, but the distributive property is used to simplify the expression. Note there is an implied one in front of the first set of parentheses and an implied -1 in front of the second set of parentheses. Distributing the 1, -1, and 2 produces:

$$1(3x) + 1(2y) + 1(1) - 1(5x) - 1(-3) + 2(3y) + 2(4)$$

$$3x + 2y + 1 - 5x + 3 + 6y + 8$$

This expression contains like terms that are combined to produce the simplified expression:

$$-2x + 8y + 12$$

Algebraic expressions are tested to be equivalent by choosing values for the variables and evaluating both expressions. For example, $4(3x - 2)$ and $12x - 8$ are tested by substituting 3 for the variable *x* and calculating to determine if equivalent values result.

## Evaluating Simple Expressions for Given Values

As stated previously, an algebraic expression is a statement written in mathematical symbols, typically including one or more unknown values represented by variables. For example, the expression $2x + 3$ states that an unknown number (*x*) is multiplied by 2 and added to 3. If given a value for the unknown number, or variable, the value of the expression is determined. For example, if the value of the variable *x* is 4, the value of the expression 4 is multiplied by 2, and 3 is added. This results in a value of 11 for the expression.

When given an algebraic expression and values for the variable(s), the expression is evaluated to determine its numerical value. To evaluate the expression, the given values for the variables are substituted (or replaced), and the expression is simplified using the order of operations. Parenthesis should be used when substituting. Consider the following: Evaluate $a - 2b + ab$ for $a = 3$ and $b = -1$. To evaluate, any variable *a* is replaced with 3 and any variable *b* with -1, producing:

$$3(-2)(-1) + (3)(-1)$$

Next, the order of operations is used to calculate the value of the expression, which is 2.

## Verbal Statements and Algebraic Expressions

As mentioned, an algebraic expression is a statement about unknown quantities expressed in mathematical symbols. The statement *five times a number added to forty* is expressed as $5x + 40$. An equation is a statement in which two expressions (with at least one containing a variable) are equal to one another. The statement *five times a number added to forty is equal to ten* is expressed as $5x + 40 = 10$.

Real world scenarios can also be expressed mathematically. Suppose a job pays its employees $300 per week and $40 for each sale made. The weekly pay is represented by the expression $40x + 300$ where *x* is the number of sales made during the week.

Consider the following scenario: Bob had $20 and Tom had $4. After selling 4 ice cream cones to Bob, Tom has as much money as Bob. The cost of an ice cream cone is an unknown quantity and can be represented by a variable (*x*). The amount of money Bob has after his purchase is four times the cost of an ice cream cone subtracted from his original $20 $\rightarrow 20 - 4x$. The amount of money Tom has after his sale is four times the cost of an ice cream cone added to his original $\$4 \rightarrow 4x + 4$. After the sale, the amount of money that Bob and Tom have are equal:

$$20 - 4x = 4x + 4$$

When expressing a verbal or written statement mathematically, it is vital to understand words or phrases that can be represented with symbols. The following are examples:

| Symbol | Phrase |
|---|---|
| + | Added to; increased by; sum of; more than |
| − | Decreased by; difference between; less than; take away |
| × | Multiplied by; 3(4,5...) times as large; product of |
| ÷ | Divided by; quotient of; half (third, etc.) of |
| = | Is; the same as; results in; as much as; equal to |
| x,t,n, etc. | A number; unknown quantity; value of; variable |

## Rational Expressions

A **rational expression** is a fraction where the numerator and denominator are both polynomials. Some examples of rational expressions include the following: $\frac{4x^3y^5}{3z^4}$, $\frac{4x^3+3x}{x^2}$, and $\frac{x^2+7x+10}{x+2}$. Since these refer to expressions and not equations, they can be simplified but not solved. Some rational expressions with monomials can be simplified. Other rational expressions such as the last example, $\frac{x^2+7x+10}{x+2}$, take more steps to be simplified. First, the polynomial on top can be factored from $x^2 + 7x + 10$ into $(x + 5)(x + 2)$. Then the common factors can be canceled and the expression can be simplified to $(x + 5)$.

Consider this problem as an example of using rational expressions. Reggie wants to lay sod in his rectangular backyard. The length of the yard is given by the expression $4x + 2$ and the width is unknown. The area of the yard is $20x + 10$. Reggie needs to find the width of the yard. Knowing that the area of a rectangle is length multiplied by width, an expression can be written to find the width: $\frac{20x+10}{4x+2}$, area divided by length. Simplifying this expression by factoring out 10 on the top and 2 on the bottom leads to this expression:

$$\frac{10(2x + 1)}{2(2x + 1)}$$

By cancelling out the $2x + 1$, that results in $\frac{10}{2} = 5$. The width of the yard is found to be 5 by simplifying a rational expression.

# Equations

As mentioned, an **equation** is a statement in which two expressions (at least one containing a variable) are equal to each other. An algebraic expression can be thought of as a mathematical phrase and an equation can be thought of as a mathematical sentence.

Linear equations and linear inequalities are both comparisons of two algebraic expressions. However, unlike equations in which the expressions are equal, linear inequalities compare expressions that may be unequal. Linear equations typically have one value for the variable that makes the statement true. Linear inequalities generally have an infinite number of values that make the statement true.

When solving a **linear equation**, the desired result requires determining a numerical value for the unknown variable. If given a linear equation involving addition, subtraction, multiplication, or division, working backwards isolates the variable. Addition and subtraction are inverse operations, as are multiplication and division. Therefore, they can be used to cancel each other out.

The first steps to solving linear equations are distributing, if necessary, and combining any like terms on the same side of the equation. Sides of an equation are separated by an **equal** sign. Next, the equation is manipulated to show the variable on one side. Whatever is done to one side of the equation must be done to the other side of the equation to remain equal. Inverse operations are then used to isolate the variable and undo the order of operations backwards. Addition and subtraction are undone, then multiplication and division are undone.

For example, solve $4(t-2)+2t-4=2(9-2t)$

Distributing: $4t-8+2t-4=18-4t$

Combining like terms: $6t-12=18-4t$

Adding $4t$ to each side to move the variable: $10t-12=18$

Adding 12 to each side to isolate the variable: $10t=30$

Dividing each side by 10 to isolate the variable: $t=3$

The answer can be checked by substituting the value for the variable into the original equation, ensuring that both sides calculate to be equal.

## Linear Equations

**Linear relationships** describe the way two quantities change with respect to each other. The relationship is defined as linear because a line is produced if all the sets of corresponding values are graphed on a coordinate grid. When expressing the linear relationship as an equation, the equation is often written in the form $y=mx+b$ (slope-intercept form) where $m$ and $b$ are numerical values and $x$ and $y$ are variables (for example, $y=5x+10$). Given a linear equation and the value of either variable ($x$ or $y$), the value of the other variable can be determined.

Suppose a teacher is grading a test containing 20 questions with 5 points given for each correct answer, adding a curve of 10 points to each test. This linear relationship can be expressed as the equation $y=5x+10$ where $x$ represents the number of correct answers, and y represents the test score. To determine the score of a test with a given number of correct answers, the number of correct answers is substituted into the equation for $x$ and evaluated. For example, for 10 correct answers, 10 is substituted for $x$:

$$y=5(10)+10 \rightarrow y=60$$

Therefore, 10 correct answers will result in a score of 60. The number of correct answers needed to obtain a certain score can also be determined. To determine the number of correct answers needed to score a 90, 90 is substituted for $y$ in the equation ($y$ represents the test score) and solved:

$$90=5x+10$$

$$80=5x \rightarrow 16=x$$

Therefore, 16 correct answers are needed to score a 90.

Linear relationships may be represented by a table of 2 corresponding values. Certain tables may determine the relationship between the values and predict other corresponding sets. Consider the table below, which displays the money in a checking account that charges a monthly fee:

| **Month** | 0 | 1 | 2 | 3 | 4 |
|---|---|---|---|---|---|
| **Balance** | $210 | $195 | $180 | $165 | $150 |

An examination of the values reveals that the account loses $15 every month (the month increases by one and the balance decreases by 15). This information can be used to predict future values. To determine what the value will be in month 6, the pattern can be continued, and it can be concluded that the balance will be $120. To determine which month the balance will be $0, $210 is divided by $15 (since the balance decreases $15 every month), resulting in month 14.

Similar to a table, a graph can display corresponding values of a linear relationship.

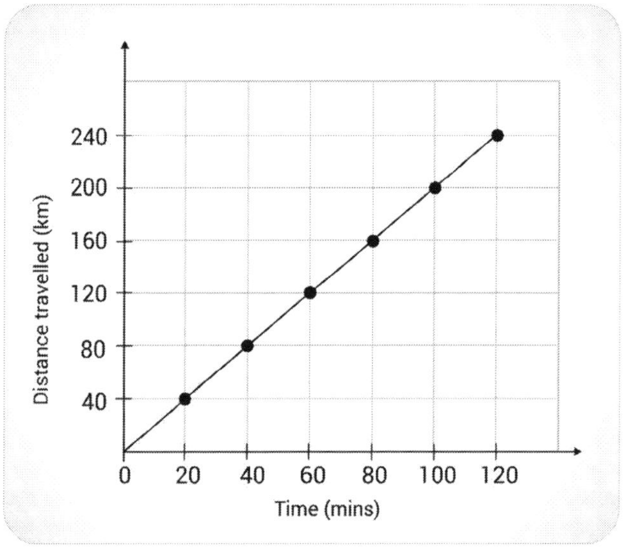


The graph above represents the relationship between distance traveled and time. To find the distance traveled in 80 minutes, the mark for 80 minutes is located at the bottom of the graph. By following this mark directly up on the graph, the corresponding point for 80 minutes is directly across from the 160 kilometer mark. This information indicates that the distance travelled in 80 minutes is 160 kilometers. To predict information not displayed on the graph, the way in which the variables change with respect to one another is determined. In this case, distance increases by 40 kilometers as time increases by 20 minutes. This information can be used to continue the data in the graph or convert the values to a table.

## Determining the Equation of a Line

An equation is called **linear** if it can take the form of the equation $y = ax + b$, for any two numbers *a* and *b*. A linear equation forms a straight line when graphed on the coordinate plane. An example of a linear equation is shown below on the graph.

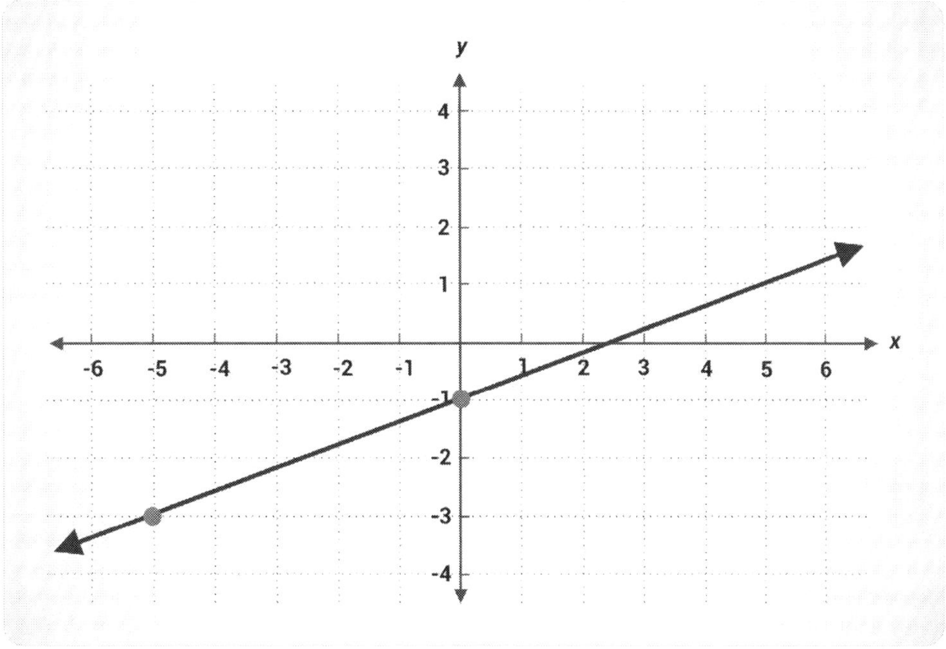


This is a graph of the following function:

$$y = \frac{2}{5}x - 1$$

A table of values that satisfies this function is shown below.

| x | y |
|---|---|
| -5 | -3 |
| 0 | -1 |
| 5 | 1 |
| 10 | 3 |

These points can be found on the graph using the form (x,y).

Given the graph of a line, its equation can be written in two ways. If the *y*-intercept is easily identified (is an integer), it and another point can be used to determine the slope. When determining $\frac{change\ in\ y}{change\ in\ x}$ from one point to another on the graph, the distance for $\frac{rise}{run}$ is being figured. The equation should be written in slope-intercept form, $y = mx + b$, with *m* representing the slope and *b* representing the *y*-intercept.

The equation of a line can also be written by identifying two points on the graph of the line. To do so, the slope is calculated and then the values are substituted for the slope and either of the ordered pairs into the point-slope form of an equation.

## The Basic Forms of Linear Equations

Linear functions may take forms other than $y = ax + b$. The most common forms of linear equations are explained below:

1. Standard Form: $Ax + By = C$, in which the slope is given by $m = \frac{-A}{B}$, and the *y*-intercept is given by $\frac{C}{B}$.

2. Slope-Intercept Form: $y = mx + b$, where the slope is *m* and the *y* intercept is *b*.

3. Point-Slope Form: $y - y_1 = m(x - x_1)$, where the slope is *m* and $(x_1, y_1)$ is any point on the chosen line.

4. Two-Point Form: $\frac{y-y_1}{x-x_1} = \frac{y_2-y_1}{x_2-x_1}$, where $(x_1, y_1)$ and $(x_2, y_2)$ are any two distinct points on the chosen line. Note that the slope is given by $m = \frac{y_2-y_1}{x_2-x_1}$.

5. Intercept Form: $\frac{x}{x_1} + \frac{y}{y_1} = 1$, in which $x_1$ is the *x*-intercept and $y_1$ is the *y*-intercept.

These five ways to write linear equations are all useful in different circumstances. Depending on the given information, it may be easier to write one of the forms over another.

If $y = mx$, *y* is directly proportional to *x*. In this case, changing *x* by a factor changes *y* by that same factor. If $y = \frac{m}{x}$, *y* is inversely proportional to *x*. For example, if *x* is increased by a factor of 3, then *y* will be decreased by the same factor, 3.

## Writing Linear Equations in Two Variables

When writing linear equations in two variables, the process depends on the information given. Questions will typically provide the slope of the line and its *y*-intercept, an ordered pair and the slope, or two ordered pairs.

### *Given the Slope and Y-Intercept*

Linear equations are commonly written in **slope-intercept form**, $y = mx + b$, where *m* represents the **slope** of the line and *b* represents the ***y*-intercept**. The slope is the rate of change between the variables, usually expressed as a whole number or fraction. The *y*-intercept is the value of *y* when *x* = 0 (the point where the line intercepts the *y*-axis on a graph). Given the slope and *y*-intercept of a line, the values are substituted for *m* and *b* into the equation. A line with a slope of ½ and *y*-intercept of -2 would have an equation $y = \frac{1}{2}x - 2$.

### *Given an Ordered Pair and the Slope*

The **point-slope** form of a line, $y - y_1 = m(x - x_1)$, is used to write an equation when given an ordered pair (point on the equation's graph) for the function and its rate of change (slope of the line). The values for the slope, *m*, and the point ($x_1$, $y_1$) are substituted into the point-slope form to obtain the equation of the line. A line with a slope of 3 and an ordered pair (4, -2) would have an equation:

$$y - (-2) = 3(x - 4)$$

If a question specifies that the equation be written in slope-intercept form, the equation should be manipulated to isolate *y*:

Solve: $y - (-2) = 3(x - 4)$

Distribute: $y + 2 = 3x - 12$

Subtract 2 from both sides: $y = 3x - 14$

### *Given Two Ordered Pairs*

Given two ordered pairs for a function, ($x_1$, $y_1$) and ($x_2$, $y_2$), it is possible to determine the rate of change between the variables (slope of the line). To calculate the slope of the line, m, the values for the ordered pairs should be substituted into the formula:

$$m = \frac{y_2 - y_1}{x_2 - x_1}$$

The expression is substituted to obtain a whole number or fraction for the slope. Once the slope is calculated, the slope and either of the ordered pairs should be substituted into the point-slope form to obtain the equation of the line.

## Linear Equations in One Variable – Special Cases

Solving a linear equation produces a value for the variable that makes the algebraic statement true. If there is no value for the variable that would make the statement true, there is no solution to the equation. Here's a sample equation:

$$x + 3 = x - 1$$

There is no value for *x* in which adding 3 to the value would produce the same result as subtracting 1 from that value. Conversely, if any value for the variable would make a true statement, the equation has an infinite number of solutions. Here's another sample equation:

$$3x + 6 = 3(x + 2)$$

Any real number substituted for *x* would result in a true statement (both sides of the equation are equal).

By manipulating equations similar to the two above, the variable of the equation will cancel out completely. If the constants that are left express a true statement (ex., $6 = 6$), then all real numbers are solutions to the equation. If the constants left express a false statement (ex., $3 = -1$), then there is no solution to the equation.

A question on this material may present a linear equation with an unknown value for either a constant or a coefficient of the variable and ask to determine the value that produces an equation with no solution or infinite solutions. For example:

> $3x + 7 = 3x + 10 + n$; Find the value of *n* that would create an equation with an infinite number of solutions for the variable *x*.

To solve this problem, the equation should be manipulated so the variable *x* will cancel. To do this, 3*x* should be subtracted from both sides, which would leave $7 = 10 + n$. By subtracting 10 on both sides, it is determined that $n = -3$. Therefore, a value of -3 for *n* would result in an equation with a solution set of all real numbers.

If the same problem asked for the equation to have no solution, the value of *n* would be all real numbers except -3.

## Solving Quadratic Equations

Solving quadratic equations is a little trickier. If they take the form $ax^2 - b = 0$, then the equation can be solved by adding *b* to both sides and dividing by *a* to get $x^2 = \frac{b}{a}$. Then, the solution is:

$$x = \pm\sqrt{\frac{b}{a}}$$

Note that this is actually two separate solutions, unless *b* happens to be zero.

If a quadratic equation has no constant—so that it takes the form $ax^2 + bx = 0$—then the *x* can be factored out to get:

$$x(ax + b) = 0$$

Then, the solutions are $x = 0$, together with the solutions to $ax + b = 0$. Both factors $x$ and $(ax + b)$ can be set equal to zero to solve for $x$ because one of those values must be zero for their product to equal zero. For an equation $ab = 0$ to be true, either $a = 0$, or $b = 0$.

A given quadratic equation $x^2 + bx + c$ can be factored into $(x + A)(x + B)$, where $A + B = b$, and $AB = c$. Finding the values of *A* and *B* can take time, but such a pair of numbers can be found by guessing and checking. Looking at the positive and negative factors for *c* offers a good starting point.

For example, in $x^2 - 5x + 6$, the factors of 6 are 1, 2, and 3. Now, $(-2)(-3) = 6$, and $-2 - 3 = -5$. In general, however, this may not work, in which case another approach may need to be used.

A quadratic equation of the form $x^2 + 2xb + b^2 = 0$ can be factored into:

$$(x + b)^2 = 0$$

Similarly, $x^2 - 2xy + y^2 = 0$ factors into $(x - y)^2 = 0$.

In general, the constant term may not be the right value to be factored this way. A more general method for solving these quadratic equations must then be found. The following two methods will work in any situation.

The first method is called **completing the square.** The idea here is that in any equation $x^2 + 2xb + c = 0$, something could be added to both sides of the equation to get the left side to look like $x^2 + 2xb + b^2$, meaning it could be factored into $(x + b)^2 = 0$.

Example: $x^2 + 6x - 1 = 0$

The left-hand side could be factored if the constant were equal to 9, since:

$$x^2 + 6x + 9 = (x + 3)^2$$

To get a constant of 9 on the left, 10 needs to be added to both sides.

That changes the equation to:

$$x^2 + 6x + 9 = 10$$

Factoring the left gives:

$$(x + 3)^2 = 10$$

Then, the square root of both sides can be taken (remembering that this introduces a $\pm$):

$$x + 3 = \pm\sqrt{10}$$

Finally, 3 is subtracted from both sides to get two solutions:

$$x = -3 \pm \sqrt{10}$$

The first method of completing the square can be used in finding the second method, the **quadratic formula.** It can be used to solve any quadratic equation. This formula may be the longest method for solving quadratic equations and is commonly used as a last resort after other methods are ruled out.

It can be helpful in memorizing the formula to see where it comes from, so here are the steps involved.

The most general form for a quadratic equation is:

$$ax^2 + bx + c = 0$$

First, dividing both sides by *a* leaves us with:

$$x^2 + \frac{b}{a}x + \frac{c}{a} = 0$$

To complete the square on the left-hand side, $\frac{c}{a}$ can be subtracted on both sides to get:

$$x^2 + \frac{b}{a}x = -\frac{c}{a}$$

$(\frac{b}{2a})^2$ is then added to both sides.

This gives:

$$x^2 + \frac{b}{a}x + (\frac{b}{2a})^2 = (\frac{b}{2a})^2 - \frac{c}{a}$$

The left can now be factored and the right-hand side simplified to give:

$$(x + \frac{b}{2a})^2 = \frac{b^2 - 4ac}{4a}$$

Taking the square roots gives:

$$x + \frac{b}{2a} = \pm\frac{\sqrt{b^2 - 4ac}}{2a}$$

Solving for x yields the quadratic formula:

$$x = \frac{-b \pm \sqrt{b^2 - 4ac}}{2a}$$

It isn't necessary to remember how to get this formula, but memorizing the formula itself is the goal.

If an equation involves taking a root, then the first step is to move the root to one side of the equation and everything else to the other side. That way, both sides can be raised to the index of the radical in order to remove it, and solving the equation can continue.

## Solving Equations with Exponents and Roots

Here are some of the most important properties of exponents and roots: if *n* is an integer, and if $a^n = b^n$, then $a = b$ if *n* is odd; but $a = \pm b$ if *n* is even. Similarly, if the roots of two things are equal, $\sqrt[n]{a} = \sqrt[n]{b}$, then $a = b$. This means that when starting with a true equation, both sides of that equation can be raised to a given power to obtain another true equation. Remember that when an even-powered root is taken on both sides of the equation, a $\pm$ in the result. For example, given the equation $x^2 = 16$, take the square root of both sides to solve for x. This results in the answer $x = \pm 4$ because $(-4)^2 = 16$ and $(4)^2 = 16$.

Another property is that if $a^n = a^m$, then $n = m$. This is true for any real numbers *n* and *m*.

For solving the equation $\sqrt{x+2} - 1 = 3$, start by moving the -1 over to the right-hand side. This is performed by adding 1 to both sides, which yields $\sqrt{x+2} = 4$. Now, square both sides, but remember that by squaring both sides, the signs are irrelevant. This yields $x + 2 = 16$, which simplifies to give $x = 14$.

Now consider the problem $(x+1)^4 = 16$. To solve this, take the fourth root of both sides, which means an ambiguity in the sign will be introduced because it is an even root:

$$\sqrt[4]{(x+1)^4} = \pm\sqrt[4]{16}$$

The right-hand side is 2, since $2^4 = 16$. Therefore, $x + 1 = \pm 2$ or $x = -1 \pm 2$. Thus, the two possible solutions are $x = -3$ and $x = 1$.

Remember that when solving equations, the answer can be checked by plugging the solution back into the problem to make a true statement.

In general, the following rules for exponents should be used for any numbers $a, b, m,$ and $n$.

- $a^1 = a$.
- $1^a = 1$.
- $a^0 = 1$.
- $a^m a^n = a^{m+n}$.
- $\frac{a^m}{a^n} = a^{m-n}$
- $(a^m)^n = a^{m \times n}$.
- $(ab)^m = a^m b^m$.
- $(\frac{a}{b})^m = \frac{a^m}{b^m}$.

As an example of applying these rules, consider the problem of simplifying the expression:

$$(3x^2y)^3(2xy^4)$$

Start by simplifying the left term using the sixth rule listed. Applying this rule yields the following expression:

$$27x^6y^3(2xy^4)$$

The exponents can now be combined with base *x* and the exponents with base *y*. Multiply the coefficients to yield:

$$54x^7y^7$$

## Expressing Systems of Two Linear Equations in Two Variables

A **system of two linear equations in two variables** is a set of equations that use the same variables, usually *x* and *y*. Here's a sample problem:

> An Internet provider charges an installation fee and a monthly charge. It advertises that two months of its offering costs $100 and six months costs $200. Find the monthly charge and the installation fee.

The two unknown quantities (variables) are the monthly charge and the installation fee. There are two different statements given relating the variables: two months added to the installation fee is $100; and six months added to the installation fee is $200. Using the variable *x* as the monthly charge and *y* as the installation fee, the statements can be written as the following: $2x + y = 100$; $6x + y = 200$. These two equations taken together form a system modeling the given scenario.

## Solutions of a System of Two Linear Equations in Two Variables

A solution for a system of equations is an ordered pair that makes both equations true. One method for solving a system of equations is to graph both lines on a coordinate plane. If the lines intersect, the point of intersection is the solution to the system. Every point on a line represents an ordered pair that makes its equation true. The ordered pair represented by this point of intersection lies on both lines and therefore makes both equations true. This ordered pair should be checked by substituting its values into both of the original equations of the system. Note that given a system of equations and an ordered pair, the ordered pair can be determined to be a solution or not by checking it in both equations.

If, when graphed, the lines representing the equations of a system do not intersect, then the two lines are parallel to each other or they are the same exact line. Parallel lines extend in the same direction without ever meeting. A system consisting of parallel lines has no solution. If the equations for a system represent the same exact line, then every point on the line is a solution to the system. In this case, there would be an infinite number of solutions. A system consisting of intersecting lines is referred to as **independent**, a system consisting of parallel lines is referred to as **inconsistent**, and a system consisting of coinciding lines is referred to as **dependent**.

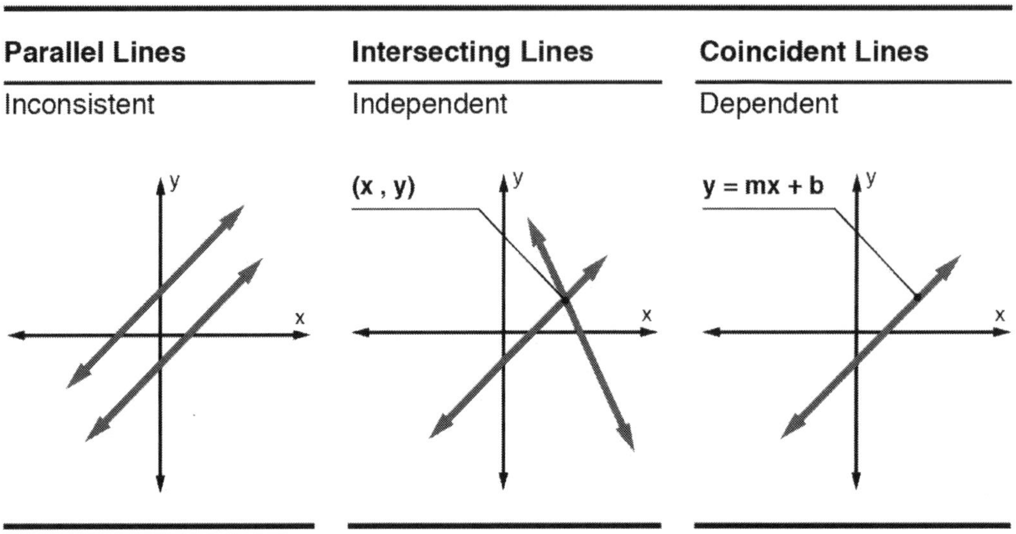


## Algebraically Solving Systems of Two Linear Equations in Two Variables

A system of two linear equations in two variables is a set of equations that use the same variables (typically *x* and *y*). A solution to the system is an ordered pair that makes both equations true. One method for solving a system is by graphing as was just explained. This method, however, is not always practical. Students may not have graph paper; or the solution may not consist of integers, making it difficult to identify the exact point of intersection on a graph. There are two methods for solving systems of equations algebraically: substitution and elimination. The method used will depend on the characteristics of the equations in the system.

### *Solving Systems of Equations with the Substitution Method*

If one of the equations in a system has an isolated variable (*x*= or *y*=) or a variable that can be easily isolated, the substitution method can be used. Here's a sample system: $x + 3y = 7; 2x + 4y = 24$. The first equation can easily be solved for *x*. By subtracting 3*y* on both sides, the resulting equation is $x = 7 - 3y$. When one equation is solved for a variable, the expression that it is equal can be substituted into the other equation. For this example, $(7 - 3y)$ would be substituted for *x* into the second equation as follows:

$$2(7 - 3y) + 4y = 24$$

Solving this equation results in $y = -5$. Once the value for one variable is known, this value should be substituted into either of the original equations to determine the value of the other variable. For the example, -5 would be substituted for *y* in either of the original equations. Substituting into the first equation results in $x + 3(-5) = 7$, and solving this equation yields $x = 22$. The solution to a system is an ordered pair, so the solution to the example is written as (22, 7). The solution can be checked by substituting it into both equations of the system to ensure it results in two true statements.

### *Solving Systems of Equations with the Elimination Method*

The elimination method for solving a system of equations involves canceling out (or eliminating) one of the variables. This method is typically used when both equations of a system are written in standard form ($Ax + By = C$). An example is $2x + 3y = 12; 5x - y = 13$. To perform the elimination method, the equations in the system should be arranged vertically to be added together and then one or both of the equations should be multiplied so that one variable will be eliminated when the two are added. Opposites will cancel each other when added together. For example, 8*x* and -8*x* will cancel each other when added. For the example above, writing the system vertically helps identify that the bottom equation should be multiplied by 3 to eliminate the variable *y*.

$$2x + 3y = 12 \quad \rightarrow \quad 2x + 3y = 12$$

$$3(5x - y = 13) \quad \rightarrow \quad 15x - 3y = 39$$

Adding the two equations together vertically results in $17x = 51$. Solving yields $x = 3$. Once the value for one variable is known, it can be substituted into either of the original equations to determine the value of the other variable. Once this is obtained, the solution can be written as an ordered pair (*x*, *y*) and checked in both equations of the system. In this example, the solution is (3, 2).

## Inequalities

**Linear inequalities** express the relationship between unequal values. More specifically, they describe in what way the values are unequal. A value can be greater than (>), less than (<), greater than or equal to (≥), or less than or equal to (≤) another value. $5x + 40 > 65$ is read as *five times a number added to forty is greater than sixty-five.*

When solving a linear inequality, the solution is the set of all numbers that make the statement true. The inequality $x + 2 \geq 6$ has a solution set of 4 and every number greater than 4 (4.01; 5; 12; 107; etc.). Adding 2 to 4 or any number greater than 4 results in a value that is greater than or equal to 6. Therefore, $x \geq 4$ is the solution set.

To algebraically solve a linear inequality, follow the same steps as those for solving a linear equation. The inequality symbol stays the same for all operations *except* when multiplying or dividing by a negative number. If multiplying or dividing by a negative number while solving an inequality, the relationship reverses (the sign flips). In other words, > switches to < and vice versa. Multiplying or dividing by a positive number does not change the relationship, so the sign stays the same. An example is shown below.

Solve $-2x - 8 \leq 22$

Add 8 to both sides: $-2x \leq 30$

Divide both sides by -2: $x \geq -15$

Solutions of a linear equation or a linear inequality are the values of the variable that make a statement true. In the case of a linear equation, the solution set (list of all possible solutions) typically consists of a single numerical value. To find the solution, the equation is solved by isolating the variable. For example, solving the equation $3x - 7 = -13$ produces the solution $x = -2$. The only value for *x* which produces a true statement is -2. This can be checked by substituting -2 into the original equation to check that both sides are equal. In this case, $3(-2) - 7 = -13 \rightarrow -13 = -13$; therefore, -2 is a solution.

Although linear equations generally have one solution, this is not always the case. If there is no value for the variable that makes the statement true, there is no solution to the equation. Consider the equation $x + 3 = x - 1$. There is no value for *x* in which adding 3 to the value produces the same result as subtracting one from the value. Conversely, if any value for the variable makes a true statement, the equation has an infinite number of solutions. Consider the equation $3x + 6 = 3(x + 2)$. Any number substituted for *x* will result in a true statement (both sides of the equation are equal).

By manipulating equations like the two above, the variable of the equation will cancel out completely. If the remaining constants express a true statement (e.g., $6 = 6$), then all real numbers are solutions to the equation. If the constants left express a false statement (e.g., $3 = -1$), then no solution exists for the equation.

Solving a linear inequality requires all values that make the statement true to be determined. For example, solving $3x - 7 \geq -13$ produces the solution $x \geq -2$. This means that -2 and any number greater than -2 produces a true statement. Solution sets for linear inequalities will often be displayed using a number line. If a value is included in the set (≥ or ≤), a shaded dot is placed on that value and an arrow extending in the direction of the solutions. For a variable > or ≥ a number, the arrow will point right on a number line, the direction where the numbers increase. If a variable is < or ≤ a number, the arrow will point left on a number line, which is the direction where the numbers decrease. If the value is not included in the set (> or <), an open (unshaded) circle on that value is used with an arrow in the appropriate direction.

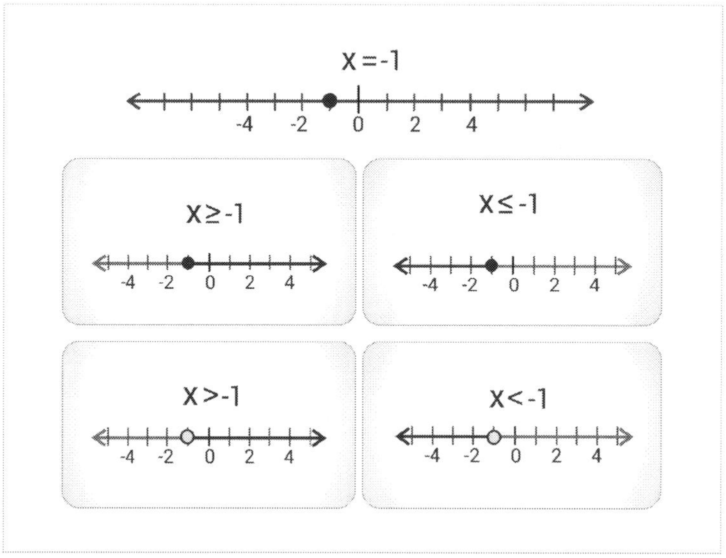


Similar to linear equations, a linear inequality may have a solution set consisting of all real numbers, or can contain no solution. When solved algebraically, a linear inequality in which the variable cancels out and results in a true statement (e.g., $7 \geq 2$) has a solution set of all real numbers. A linear inequality in which the variable cancels out and results in a false statement (e.g., $7 \leq 2$) has no solution.

## Compound Inequalities

A compound inequality is a pair of inequalities joined by *and* or *or*. Given a compound inequality, to determine its solution set, both inequalities should be solved for the given variable. The solution set for a compound inequality containing *and* consists of all the values for the variable that make both inequalities true. If solving the compound inequality results in $x \geq -2$ and $x < 3$, the solution set would consist of all values between -2 and 3, including -2. This may also be written as follows: $-2 \leq x < 3$. Due to the graphs of their solution sets (shown below), compound inequalities such as these are referred to as conjunctions.

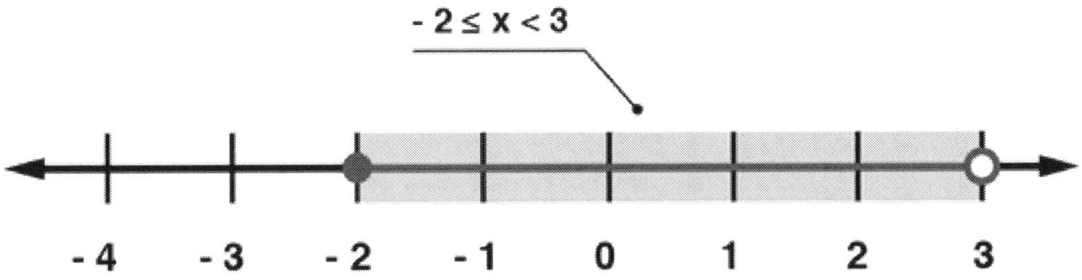


If there are no values that would make both inequalities of a compound inequality containing *and* true, then there is no solution. An example would be $x > 2$ and $x \leq 0$.

The solution set for a compound inequality containing *or* consists of all the values for the variable that make at least one of the inequalities true. The solution set for the compound inequality $x < 3$ or $x \geq 6$ consists of all values less than 3, 6, and all values greater than 6. Due to the graphs of their solution sets (shown below), compound inequalities such as these are referred to as disjunctions.

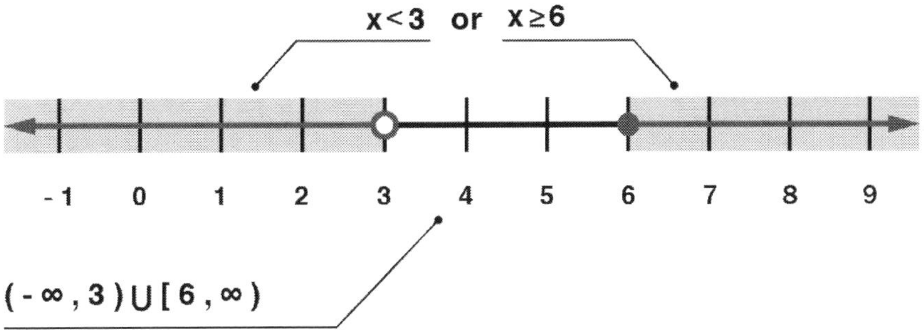


If the two inequalities for a compound inequality containing *or* "overlap," then the solution set contains all real numbers. An example would be $x > 2$ or $x < 7$. Any number would make at least one of these true.

## Expressing Linear Inequalities in Two Variables

A linear inequality in two variables is a statement expressing an unequal relationship between those two variables. Typically written in slope-intercept form, the variable *y* can be greater than; less than; greater than or equal to; or less than or equal to a linear expression including the variable *x*. Examples include $y > 3x$ and $y \leq \frac{1}{2}x - 3$. Questions may instruct students to model real world scenarios such as:

> You work part-time cutting lawns for $15 each and cleaning houses for $25 each. Your goal is to make more than $90 this week. Write an inequality to represent the possible pairs of lawns and houses needed to reach your goal.

This scenario can be expressed as $15x + 25y > 90$ where *x* is the number of lawns cut and *y* is the number of houses cleaned. As another example:

> A consumer study calls for at least 60 adult participants. It cannot use more than 25 men. Express these constraints as a system of inequalities.

This can be modeled by the system: $x + y \geq 60; x \leq 25$, where *x* represents the number of men and *y* represents the number of women. A solution to the system is an ordered pair that makes both inequalities true when substituting the values for *x* and *y*.

## Graphing Solution Sets for Systems of Linear Inequalities in Two Variables

The solution set for a system of inequalities is the region of a graph consisting of ordered pairs that make both inequalities true. To graph the solution set, each linear inequality should first be graphed with appropriate shading. The region of the graph should be identified where the shading for the two inequalities overlaps. This region contains the solution set for the system.

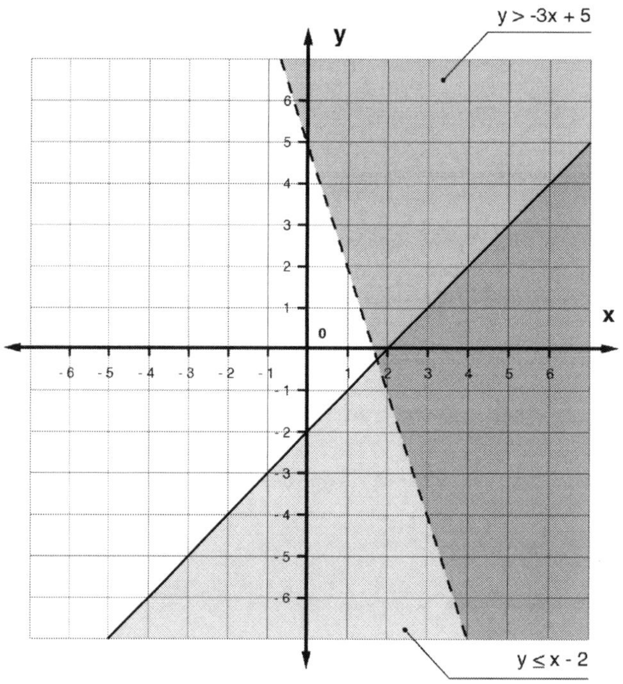


An ordered pair from the region of solutions can be selected to test in the system of inequalities.

Just as with manipulating linear inequalities in one variable, if dividing by a negative number in working with a linear inequality in two variables, the relationship reverses and the inequality sign should be flipped.

## Representation and Modeling

### Constructing, Using, and Interpreting Simple Diagrams to Solve Problems

In solving multi-step problems, the first step is to line up the available information. Then, try to decide what information the problem is asking to be found. Sketching out a diagram to illustrate what is known and what is unknown can be helpful in tackling the problem and determining what mathematical equations or processes need to be used to find the solution.

For example, you can construct a strip diagram to display the known information along with any information to be calculated. Finally, the missing information can be represented by a variable in a mathematical equation that can be solved.

For example, Delilah collects stickers and her friends gave her some stickers to add to her current collection. Joe gave her 45 stickers, and Aimee gave her 2 times the number of stickers that Joe gave Delilah. How many stickers did Delilah have to start with, if after her friends gave her more stickers, she had a total of 187 stickers?

In order to solve this, the given information must first be sorted out. Joe gives Delilah 45 stickers, Aimee gives Delilah 2 times the number Joe gives (2 × 45), and the end total of stickers is 187.

A strip diagram represents these numbers as follows:

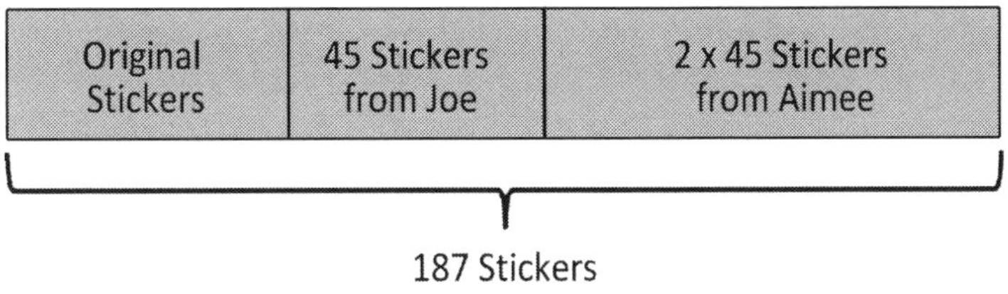


The entire situation can be modeled by this equation, using the variable *s* to stand for the original number of stickers:

$$s + 45 + (2 \times 45) = 187.$$

Solving for *s* would give the solution, as follows:

$$s + 45 + 90 = 187$$

$$s + 135 = 187$$

$$s + 135 - 135 = 187 - 135$$

$$s = 52 \text{ stickers.}$$

### Applying a Given Algorithm to Solve a Problem

When attempting to solve a math problem, it's important to apply the correct algorithm. It is much more difficult to determine what algorithm is necessary when solving word problems, because the necessary operations and equations are typically not provided. In these instances, the test taker must translate the words in the problem into true mathematical statements that can be solved.

Word problems can appear daunting, but don't let the wording intimidate you. No matter the scenario or specifics, the key to answering them is to translate the words into a math problem. Always keep in mind what the question is asking and what operations could lead to that answer. The following word problems highlight the most commonly tested question types.

Walter's Coffee Shop sells a variety of drinks and breakfast treats.

| **Price List** | |
|---|---|
| Hot Coffee | $2.00 |
| Slow-Drip Iced Coffee | $3.00 |
| Latte | $4.00 |
| Muffin | $2.00 |
| Crepe | $4.00 |
| Egg Sandwich | $5.00 |

| **Costs** | |
|---|---|
| Hot Coffee | $0.25 |
| Slow-Drip Iced Coffee | $0.75 |
| Latte | $1.00 |
| Muffin | $1.00 |
| Crepe | $2.00 |
| Egg Sandwich | $3.00 |

Walter's utilities, rent, and labor costs him $500 per day. Today, Walter sold 200 hot coffees, 100 slow-drip iced coffees, 50 lattes, 75 muffins, 45 crepes, and 60 egg sandwiches. What was Walter's total profit today?

To accurately answer this type of question, determine the total cost of making his drinks and treats, then determine how much revenue he earned from selling those products. After arriving at these two totals, the profit is measured by deducting the total cost from the total revenue.

Walter's costs for today:

| **Item** | **Quantity** | **Cost Per Unit** | **Total Cost** |
|---|---|---|---|
| Hot Coffee | 200 | $0.25 | $50 |
| Slow-Drip Iced Coffee | 100 | $0.75 | $75 |
| Latte | 50 | $1.00 | $50 |
| Muffin | 75 | $1.00 | $75 |
| Crepe | 45 | $2.00 | $90 |
| Egg Sandwich | 60 | $3.00 | $180 |
| Utilities, rent, and labor | | | $500 |
| Total Costs | | | $1,020 |

Walter's revenue for today:

| **Item** | **Quantity** | **Revenue Per Unit** | **Total Revenue** |
|---|---|---|---|
| Hot Coffee | 200 | $2.00 | $400 |
| Slow-Drip Iced Coffee | 100 | $3.00 | $300 |
| Latte | 50 | $4.00 | $200 |
| Muffin | 75 | $2.00 | $150 |
| Crepe | 45 | $4.00 | $180 |
| Egg Sandwich | 60 | $5.00 | $300 |
| Total Revenue | | | $1,530 |

Walter's Profit = *Revenue* – *Costs* = $1,530 – $1,020 = $510

This strategy is applicable to other question types. For example, calculating salary after deductions, balancing a checkbook, and calculating a dinner bill are common word problems similar to business planning. Just remember to use the correct operations. When a balance is increased, use addition. When a balance is decreased, use subtraction. Common sense and organization are your greatest assets when answering word problems.

In general, when solving word problems (also called story problems), it's important to understand what is being asked and to properly set up the initial equation. Always read the entire problem through, and then separate out what information is given in the statement. Decide what you are being asked to find and label each quantity with a variable or constant. Then write an equation to determine the unknown variable. Remember to label answers; sometimes knowing what the answers' units can help eliminate other possible solutions.

When trying to solve any word problem, look for a series of key words indicating addition, subtraction, multiplication, or division to help you determine how to set up the problem:

**Addition**: add, altogether, together, plus, increased by, more than, in all, sum, and total

**Subtraction**: minus, less than, difference, decreased by, fewer than, remain, and take away

**Multiplication**: *times*, *twice*, *of*, *double*, and *triple*

**Division**: divided by, cut up, half, quotient of, split, and shared equally

If a question asks to give words to a mathematical expression and says "equals," then an = sign must be included in the answer. Similarly, "less than or equal to" is expressed by the inequality symbol ≤, and "greater than or equal" to is expressed as ≥. Furthermore, "less than" is represented by <, and "greater than" is expressed by >.

Example:

A store is having a spring sale, where everything is 70% off. You have $45.00 to spend. A jacket is regularly priced at $80.00. Do you have enough to buy the jacket and a pair of gloves, regularly priced at $20.00?

There are two ways to approach this.

Method 1:

Set up the equations to find the sale prices: the original price minus the amount discounted.
$80.00 - ($80.00 (0.70)) = sale cost of the jacket.
$20.00 – ($20.00 (0.70)) = sale cost of the gloves.
Solve for the sale cost.
$24.00 = sale cost of the jacket.
$6.00 = sale cost of the gloves.
Determine if you have enough money for both.
$24.00 + $6.00 = total sale cost.
$30.00 is less than $45.00, so you can afford to purchase both.

Method 2:

Determine the percent of the original price that you will pay.
100% – 70% = 30%
Set up the equations to find the sale prices.
$80.00 (0.30) = cost of the jacket.
$20.00 (0.30) = cost of the gloves.
Solve.
$24.00 = cost of the jacket.
$6.00 = cost of the gloves.
Determine if you have enough money for both.
$24.00 + $6.00 = total sale cost.
$30.00 is less than $45.00, so you can afford to purchase both.

Example:

Mary and Dottie team up to mow neighborhood lawns. If Mary mows 2 lawns per hour and the two of them can mow 17.5 lawns in 5 hours, how many lawns does Dottie mow per hour?

Given rate for Mary.

$$Mary = \frac{2\ lawns}{1\ hour}$$

Unknown rate of D for Dottie.

$$Dottie = \frac{D\ lawns}{1\ hour}$$

Given rate for both.

$$Total\ mowed\ together = \frac{17.5\ lawns}{5\ hours}$$

Set up the equation for what is being asked.

$$Mary + Dottie = total\ together.$$

Fill in the givens.

$$2 + D = \frac{17.5}{5}$$

Divide.

$$2 + D = 3.5$$

Subtract 2 from both sides to isolate the variable.

$$2 - 2 + D = 3.5 - 2$$

Solve and label Dottie's mowing rate.

$$D = 1.5 \text{ lawns per hour}$$

## Concrete Models

Concrete objects are used to develop a tangible understanding of operations of rational numbers. Tools such as tiles, blocks, beads, and hundred charts are used to model problems. For example, a hundred chart ($10 \times 10$) and beads can be used to model multiplication. If multiplying 5 by 4, beads are placed across 5 rows and down 4 columns producing a product of 20. Similarly, tiles can be used to model division by splitting the total into equal groups. If dividing 12 by 4, 12 tiles are placed one at a time into 4 groups. The result is 4 groups of 3. This is also an effective method for visualizing the concept of remainders.

Representations of objects can be used to expand on the concrete models of operations. Pictures, dots, and tallies can help model these concepts. Utilizing concrete models and representations creates a foundation upon which to build an abstract understanding of the operations.

## Rational Numbers on a Number Line

A number line typically consists of integers (...3, 2, 1, 0, -1, -2, -3...), and is used to visually represent the value of a rational number. Each rational number has a distinct position on the line determined by comparing its value with the displayed values on the line. For example, if plotting -1.5 on the number line below, it is necessary to recognize that the value of -1.5 is .5 less than -1 and .5 greater than -2. Therefore, -1.5 is plotted halfway between -1 and -2.

Number lines can also be useful for visualizing sums and differences of rational numbers. Adding a value indicates moving to the right (values increase to the right), and subtracting a value indicates moving to the left (numbers decrease to the left). For example, $-3 - 2$ is displayed by starting at -3 and moving to the left 5 spaces, if the number line is in increments of 1. This will result in an answer of -5.

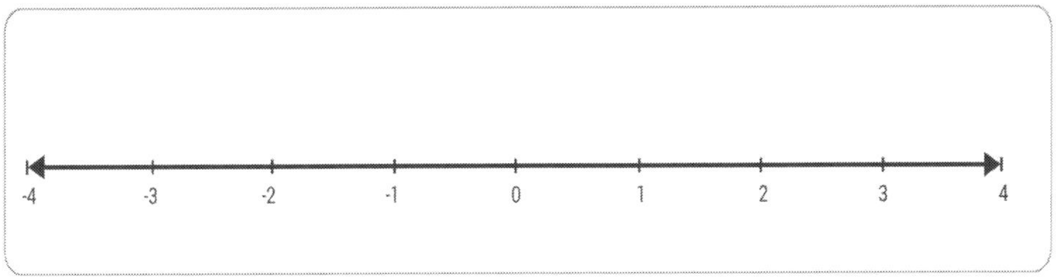

## Rectangular Arrays and Area Models

**Rectangular arrays** include an arrangement of rows and columns that correspond to the factors and display product totals.

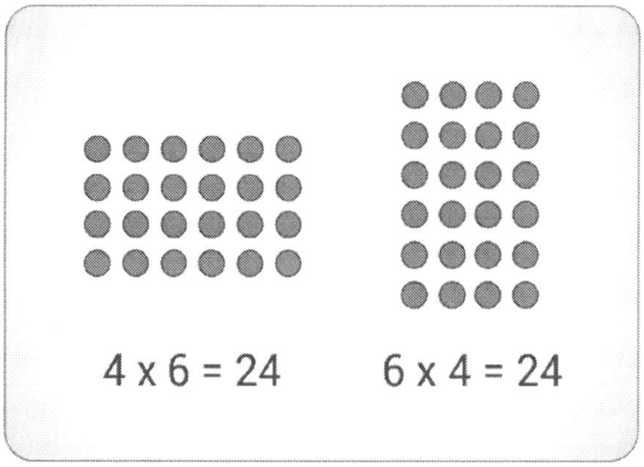


An **area model** is a rectangle that is divided into rows and columns that match up to the number of place values within each number. Take the example $29 \times 65$. These two numbers can be split into simpler numbers: $29 = 25 \times 4$ and $65 = 60 + 5$. The products of those 4 numbers are found within the rectangle and then summed up to get the answer. The entire process is: $(60 \times 25) + (60 \times 4) + (5 \times 25) + (5 \times 4) = 1{,}500 + 240 + 125 + 20 = 1{,}885$.

Here is the actual area model:

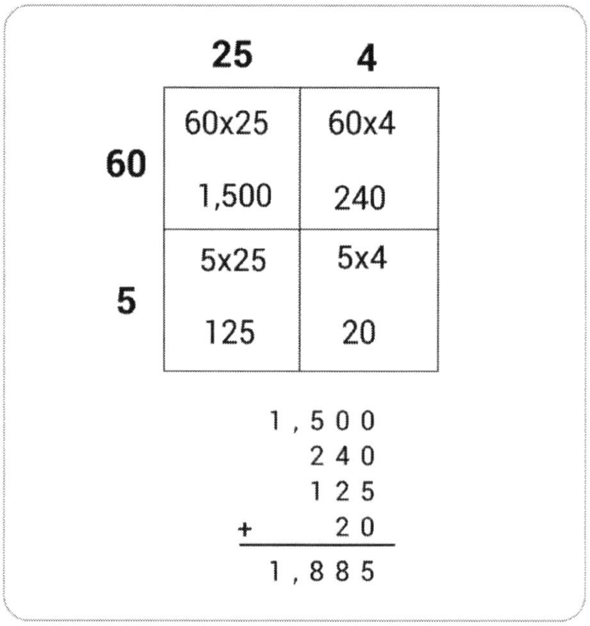


## Properties of Functions

First, it's important to understand the definition of a **relation**. Given two variables, *x* and *y*, which stand for unknown numbers, a relation between *x* and *y* is an object that splits all of the pairs (*x*, *y*) into those for which the relation is true and those for which it is false.

For example, consider the relation of $x^2 = y^2$. This relationship is true for the pair (1, 1) and for the pair (-2, 2), but false for (2, 3). Another example of a relation is $x \leq y$. This is true whenever *x* is less than or equal to *y*.

A **function** is a special kind of relation where, for each value of *x*, there is only a single value of *y* that satisfies the relation. So, $x^2 = y^2$ is *not* a function because in this case, if *x* is 1, *y* can be either 1 or -1: the pair (1, 1) and (1, -1) both satisfy the relation. More generally, for this relation, any pair of the form $(a, \pm a)$ will satisfy it. On the other hand, consider the following relation:

$$y = x^2 + 1$$

This is a function because for each value of *x*, there is a unique value of *y* that satisfies the relation. Notice, however, there are multiple values of *x* that give us the same value of *y*. This is perfectly acceptable for a function. Therefore, *y* is a function of *x*.

To determine if a relation is a function, check to see if every *x* value has a unique corresponding *y* value.

Because a function is defined as a relationship between inputs and outputs where there is only one output value for a given input, it can be viewed as an object that has *x* as its input and outputs a unique *y*-value. It is sometimes convenient to express this using **function notation**, where the function itself is given a name, often *f*. To emphasize that *f* takes *x* as its input, the function is written as $f(x)$. In the above example, the equation could be rewritten as:

$$f(x) = x^2 + 1$$

To write the value that a function yields for some specific value of *x*, that value is put in place of *x* in the function notation. For example, $f(3)$ means the value that the function outputs when the input value is 3. If $f(x) = x^2 + 1$, then:

$$f(3) = 3^2 + 1 = 10$$

Another example of a function would be $f(x) = 4x + 4$, read "*f* of *x* is equal to four times *x* plus four." In this example, the input would be *x* and the output would be f(*x*). Ordered pairs would be represented as (*x*, f(*x*)). To find the output for an input value of 3, 3 would be substituted for *x* into the function as follows: $f(3) = 4(3) + 4$, resulting in $f(3) = 16$. Therefore, the ordered pair:

$$(3, f(3)) = (3, 16)$$

Note f(*x*) is a function of *x* denoted by *f*. Functions of *x* could be named *g*(*x*), read "*g* of *x*"; *p*(*x*), read "*p* of *x*"; etc.

As an example, the following function is in function notation:

$$f(x) = 3x - 4$$

The $f(x)$ represents the output value for an input of $x$. If $x = 2$, the equation becomes:

$$f(2) = 3(2) - 4 = 6 - 4 = 2$$

The input of 2 yields an output of 2, forming the ordered pair $(2, 2)$. The following set of ordered pairs corresponds to the given function: $(2, 2), (0, -4), (-2, -10)$. The set of all possible inputs of a function is its **domain**, and all possible outputs is called the **range**. By definition, each member of the domain is paired with only one member of the range.

Functions can also be defined recursively. In this form, they are not defined explicitly in terms of variables. Instead, they are defined using previously-evaluated function outputs, starting with either $f(0)$ or $f(1)$. An example of a recursively-defined function is:

$$f(1) = 2, f(n) = 2f(n-1) + 2n, n > 1$$

The domain of this function is the set of all integers.

A function can also be viewed as a table of pairs (*x*, *y*), which lists the value for *y* for each possible value of *x*.

To evaluate functions, plug in the given value everywhere the variable appears in the expression for the function. For example, find $g(-2)$ where:

$$g(x) = 2x^2 - \frac{4}{x}$$

To complete the problem, plug in -2 in the following way:

$$g(-2) = 2(-2)^2 - \frac{4}{-2}$$

$$2 \times 4 + 2 = 8 + 2 = 10$$

## Linear Functions

Linear relationships between two quantities can be expressed in two ways: function notation or as a linear equation with two variables. The relationship is referred to as linear because its graph is represented by a line. For a relationship to be linear, both variables must be raised to the first power only.

### *Identifying Variables for Linear Models*

Linear functions, also written as linear equations in two variables, can be written to model real-world scenarios. Questions on this material will provide information about a scenario and then request a linear equation to represent the scenario. The algebraic process for writing the equation will depend on the given information. The key to writing linear models is to decipher the information given to determine what it represents in the context of a linear equation (variables, slope, ordered pairs, etc.).

The first step to writing a linear model is to identify what the variables represent. A **variable** represents an unknown quantity, and in the case of a linear equation, a specific relationship exists between the two variables (usually *x* and *y*). Within a given scenario, the variables are the two quantities that are changing. The variable *x* is considered the **independent variable** and represents the inputs of a function.

The variable *y* is considered the **dependent variable** and represents the outputs of a function. For example, if a scenario describes distance traveled and time traveled, distance would be represented by *y* and time represented by *x*. The distance traveled depends on the time spent traveling (time is independent). If a scenario describes the cost of a cab ride and the distance traveled, the cost would be represented by *y* and the distance represented by *x*. The cost of a cab ride depends on the distance traveled.

### *Identifying the Slope and Y-Intercept for Linear Models*

**Rate of change** for any line calculates the steepness of the line over a given interval. Rate of change is also known as the **slope** or rise/run. The slope of a linear function is given by the change in *y* divided by the change in $x$. So, the formula looks like this:

$$slope = \frac{y_2 - y_1}{x_2 - x_1}$$

In the graph below, two points are plotted. The first has the coordinates of (0,1), and the second point is (2,3). Remember that the $x$-coordinate is always placed first in coordinate pairs. Work from left to right when identifying coordinates. Thus, the point on the left is point 1 (0,1), and the point on the right is point 2 (2,3).

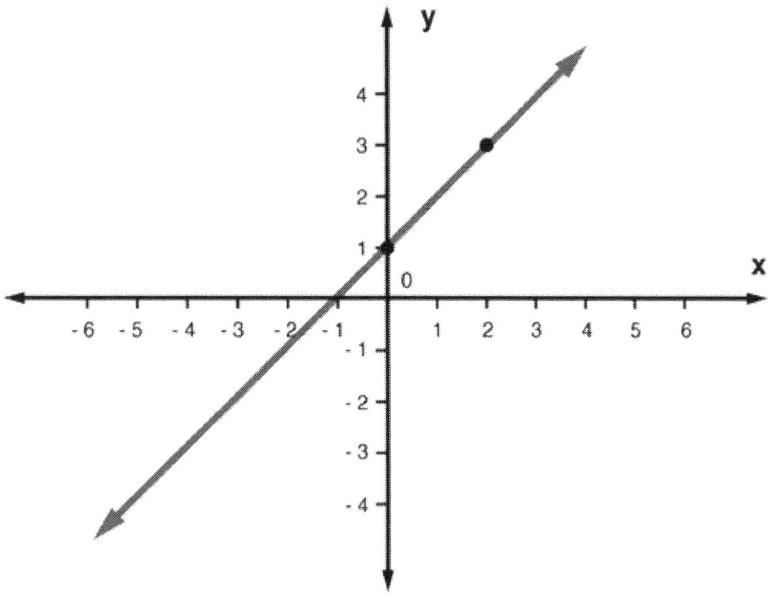


Now, we need to just plug those numbers into the equation:

$$slope = \frac{3 - 1}{2 - 0}$$

$$slope = \frac{2}{2}$$

$$slope = 1$$

This means that for every increase of 1 for $x$, $y$ also increased by 1. You can see this in the line. When $x$ is 0, $y$ is 1, and when $x$ is increased to 1, $y$ is 2.

Slope can be thought of as determining the rise over run:

$$slope = \frac{rise}{run}$$

The rise being the change vertically on the y axis and the run being the change horizontally on the $x$-axis.

Essentially, the slope of the graph of a line represents the rate of change between the variables of an equation. In the context of a real-world scenario, the slope will tell the way in which the unknown quantities (variables) change with respect to each other. A scenario involving distance and time might state that someone is traveling at a rate of 45 miles per hour. The slope of the linear model would be 45. A scenario involving the cost of a cab ride and distance traveled might state that the person is charged $3 for each mile. The slope of the linear model would be 3.

The $x$- and $y$-intercepts of a given linear function can be determined based on the form that the linear function is presented in. Of course, if the function is graphed, the $x$-intercept is the point where the graph of the line crosses the $x$-axis, while the $y$-intercept is the point where the graph of the line crosses the $y$-axis.

It is easiest to determine the intercepts of a linear function when it is written in intercept form $(\frac{x}{x_1} + \frac{y}{y_1} = 1)$. Here, $x_1$ is the $x$-intercept and $y_1$ is the $y$-intercept. When a linear function is written in standard form $(Ax + By = C)$, the y-intercept is given by $\frac{C}{B}$. The $y$-intercept of a linear equations written in slope-intercept form $(y = mx + b)$ is *b*.

The $y$-intercept of a linear function is the value of *y* when $x = 0$ (the point where the line intercepts the $y$-axis on the graph of the equation). It is sometimes helpful to think of this as a "starting point" for a linear function. Suppose for the scenario about the cab ride that the person is told that the cab company charges a flat fee of $5 plus $3 for each mile. Before traveling any distance $(x = 0)$, the cost is $5. The *y*-intercept for the linear model would be 5.

## *Identifying Ordered Pairs for Linear Models*

A linear equation with two variables can be written given a point (ordered pair) and the slope or given two points on a line. An ordered pair gives a set of corresponding values for the two variables (*x* and *y*). As an example, for a scenario involving distance and time, it is given that the person traveled 112.5 miles in 2 ½ hours. Knowing that $x$ represents time and $y$ represents distance, this information can be written as the ordered pair (2.5, 112.5).

## *Graphing a Linear Function*

The process for graphing a line depends on the form in which its equation is written: slope-intercept form or standard form.

As mentioned, when an equation is written in slope-intercept form, $y = mx + b$, *m* represents the slope of the line and *b* represents the $y$-intercept. The $y$-intercept is the value of *y* when $x = 0$ and the point at which the graph of the line crosses the *y*-axis. The slope is the rate of change between the variables, expressed as a fraction. The fraction expresses the change in *y* compared to the change in *x*. If the slope is an integer, it should be written as a fraction with a denominator of 1. For example, 5 would be written as 5/1.

To graph a line given an equation in slope-intercept form, the *y*-intercept should first be plotted. For example, to graph $y = -\frac{2}{3}x + 7$, the *y*-intercept of 7 would be plotted on the *y*-axis (vertical axis) at the point (0, 7). Next, the slope would be used to determine a second point for the line. Note that all that is necessary to graph a line is two points on that line.) The slope will indicate how to get from one point on the line to another. The slope expresses vertical change (*y*) compared to horizontal change (*x*) and therefore is sometimes referred to as $\frac{rise}{run}$. The numerator indicates the change in the *y*-value (move up for positive integers and move down for negative integers), and the denominator indicates the change in the *x* value. For the previous example, using the slope of $-\frac{2}{3}$, from the first point at the *y*-intercept, the second point should be found by counting down 2 and to the right 3. This point would be located at (3, 5).

When an equation is written in standard form, $Ax + By = C$, it is easy to identify the *x*- and *y*-intercepts for the graph of the line. Just as the *y*-intercept is the point at which the line intercepts the *y*-axis, the *x*-intercept is the point at which the line intercepts the *x*-axis. At the *y*-intercept, $x = 0$; and at the *x*-intercept, $y = 0$. Given an equation in standard form, $x = 0$ should be used to find the *y*-intercept. Likewise, $y = 0$ should be used to find the *x*-intercept. For example, to graph $3x + 2y = 6$, 0 for *y* results in $3x + 2(0) = 6$. Solving for *y* yields $x = 2$; therefore, an ordered pair for the line is (2, 0). Substituting 0 for *x* results in $3(0) + 2y = 6$. Solving for *y* yields $y = 3$; therefore, an ordered pair for the line is (0, 3). The two ordered pairs (the *x*- and *y*-intercepts) can be plotted, and a straight line through them can be constructed.

**T - chart**

| x | y |
|---|---|
| 0 | 3 |
| 2 | 0 |

**Intercepts**

**x** - intercept : (2,0)

**y** - intercept : (0,3)

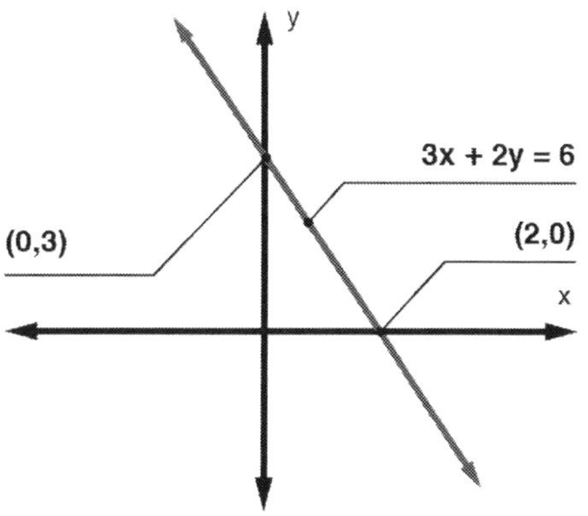


## Quadratic Functions

A **quadratic function** can be written in the standard form:

$$y = ax^2 + bx + c$$

It can be represented by a U-shaped graph called a **parabola**. The graph can either open up or open down (upside down U). The graph is symmetric about a vertical line, called the **axis of symmetry**. Corresponding points on the parabola are directly across from each other (same *y*-value) and are the same distance from the axis of symmetry (on either side). The axis of symmetry intersects the parabola at its **vertex**. For a quadratic function where the value of *a* is positive, as the inputs increase, the outputs increase until a certain value (maximum of the function) is reached. As inputs increase past the value that corresponds with the maximum output, the relationship reverses and the outputs decrease. For a quadratic function where *a* is negative, as the inputs increase, the outputs (1) decrease, (2) reach a maximum, and (3) then increase.

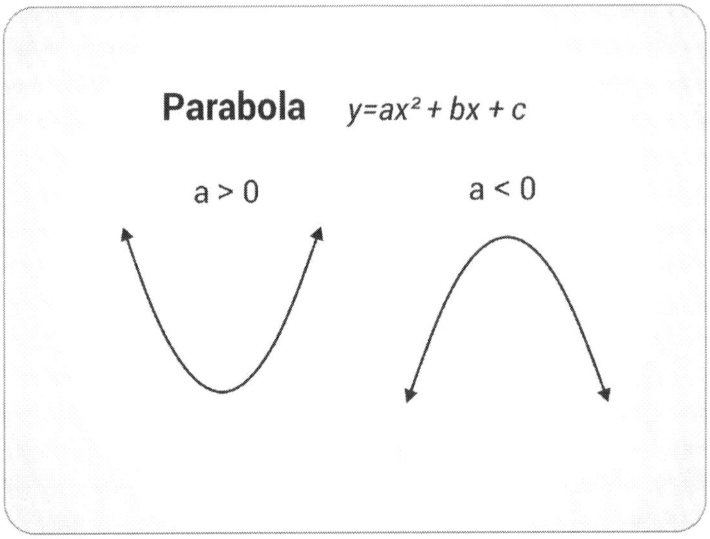


Consider a ball thrown straight up into the air. As time passes, the height of the ball increases until it reaches its maximum height. After reaching the maximum height, as time increases, the height of the ball decreases (it is falling toward the ground). This relationship can be expressed as a quadratic function where time is the input (*x*), and the height of the ball is the output (*y*).

Equations with one variable (linear equations) can be solved using the addition principle and multiplication principle. If $a = b$, then $a + c = b + c$, and $ac = bc$. Given the equation $2x - 3 = 5x + 7$, the first step is to combine the variable terms and the constant terms. Using the principles, expressions can be added and subtracted onto and off both sides of the equals sign, so the equation turns into $-10 = 3x$. Dividing by 3 on both sides through the multiplication principle with $c = \frac{1}{3}$ results in the final answer of $x = \frac{-10}{3}$.

However, this same process cannot be used to solve nonlinear equations, including quadratic equations. Quadratic equations have a higher degree than linear ones (2 versus 1) and are not solved by simply using opposite operations. When an equation has a degree of 2, completing the square is an option. For example, the quadratic equation $x^2 - 6x + 2 = 0$ can be rewritten by completing the square. The goal of completing the square is to get the equation into the form $(x - p)^2 = q$. Using the example, the constant term 2 first needs to be moved over to the opposite side by subtracting. Then, the square can be completed by adding 9 to both sides, which is the square of half of the coefficient of the middle term $-6x$. The current equation is $x^2 - 6x + 9 = 7$. The left side can be factored into a square of a binomial, resulting in $(x-3)^2 = 7$. To solve for $x$, the square root of both sides should be taken, resulting in$(x-3) = \pm\sqrt{7}$, and $x = 3 \pm \sqrt{7}$.

Other ways of solving quadratic equations include graphing, factoring, and using the quadratic formula. The equation $y = x^2 - 4x + 3$ can be graphed on the coordinate plane, and the solutions can be observed where it crosses the x-axis. The graph will be a parabola that opens up with two solutions at 1 and 3.

If quadratic equations take the form $ax^2 - b = 0$, then the equation can be solved by adding *b* to both sides and dividing by *a* to get $x^2 = \frac{b}{a}$ or $x = \pm\sqrt{\frac{b}{a}}$. Note that this is actually two separate solutions, unless *b* happens to be 0.

If a quadratic equation has no constant—so that it takes the form $ax^2 + bx = 0$—then the *x* can be factored out to get $x(ax + b) = 0$. Then, the solutions are $x = 0$, together with the solutions to $ax + b = 0$. Both factors $x$ $and$ $(ax + b)$ can be set equal to zero to solve for $x$ because one of those values must be zero for their product to equal zero. For an equation $ab = 0$ to be true, either $a = 0$, or $b = 0$.

A given quadratic equation $x^2 + bx + c$ can be factored into $(x + A)(x + B)$, where $A + B = b$, and $AB = c$. Finding the values of A and B can take time, but such a pair of numbers can be found by guessing and checking. Looking at the positive and negative factors for *c* offers a good starting point.

For example, in $x^2 - 5x + 6$, the factors of 6 are 1, 2, and 3. Now,$(-2)(-3) = 6$, and $-2 - 3 = -5$. In general, however, this may not work, in which case another approach may need to be used.

A quadratic equation of the form $x^2 + 2xb + b^2 = 0$ can be factored into $(x + b)^2 = 0$. Similarly, $x^2 - 2xy + y^2 = 0$ factors into $(x - y)^2 = 0$.

The first method of completing the square can be used in finding the second method, the quadratic formula. It can be used to solve any quadratic equation. This formula may be the longest method for solving quadratic equations and is commonly used as a last resort after other methods are ruled out.

It can be helpful in memorizing the formula to see where it comes from, so here are the steps involved.

The most general form for a quadratic equation is:

$$ax^2 + bx + c = 0$$

First, dividing both sides by *a* leaves us with:

$$x^2 + \frac{b}{a}x + \frac{c}{a} = 0$$

To complete the square on the left-hand side, c/a can be subtracted on both sides to get:

$$x^2 + \frac{b}{a}x = -\frac{c}{a}$$

$(\frac{b}{2a})^2$ is then added to both sides.

This gives:

$$x^2 + \frac{b}{a}x + (\frac{b}{2a})^2 = (\frac{b}{2a})^2 - \frac{c}{a}$$

The left can now be factored and the right-hand side simplified to give:

$$(x + \frac{b}{2a})^2 = \frac{b^2 - 4ac}{4a}$$

Taking the square roots gives:

$$x + \frac{b}{2a} = \pm\frac{\sqrt{b^2 - 4ac}}{2a}$$

Solving for x yields the quadratic formula:

$$x = \frac{-b \pm \sqrt{b^2 - 4ac}}{2a}$$

It isn't necessary to remember how to get this formula, but memorizing the formula itself is the goal.

If an equation involves taking a root, then the first step is to move the root to one side of the equation and everything else to the other side. That way, both sides can be raised to the index of the radical in order to remove it, and solving the equation can continue.

### *Graphing a Quadratic Function*

To graph a parabola, its vertex (high or low point for the curve) and at least two points on each side of the axis of symmetry need to be determined. Given a quadratic function in standard form, the axis of symmetry for its graph is the line $x = -\frac{b}{2a}$. The vertex for the parabola has an *x*-coordinate of $-\frac{b}{2a}$.

To find the *y*-coordinate for the vertex, the calculated *x*-coordinate needs to be substituted. To complete the graph, two different *x*-values need to be selected and substituted into the quadratic function to obtain the corresponding *y*-values. This will give two points on the parabola. These two points and the axis of symmetry are used to determine the two points corresponding to these. The corresponding points are the same distance from the axis of symmetry (on the other side) and contain the same *y*-coordinate.

Plotting the vertex and four other points on the parabola allows for constructing the curve.

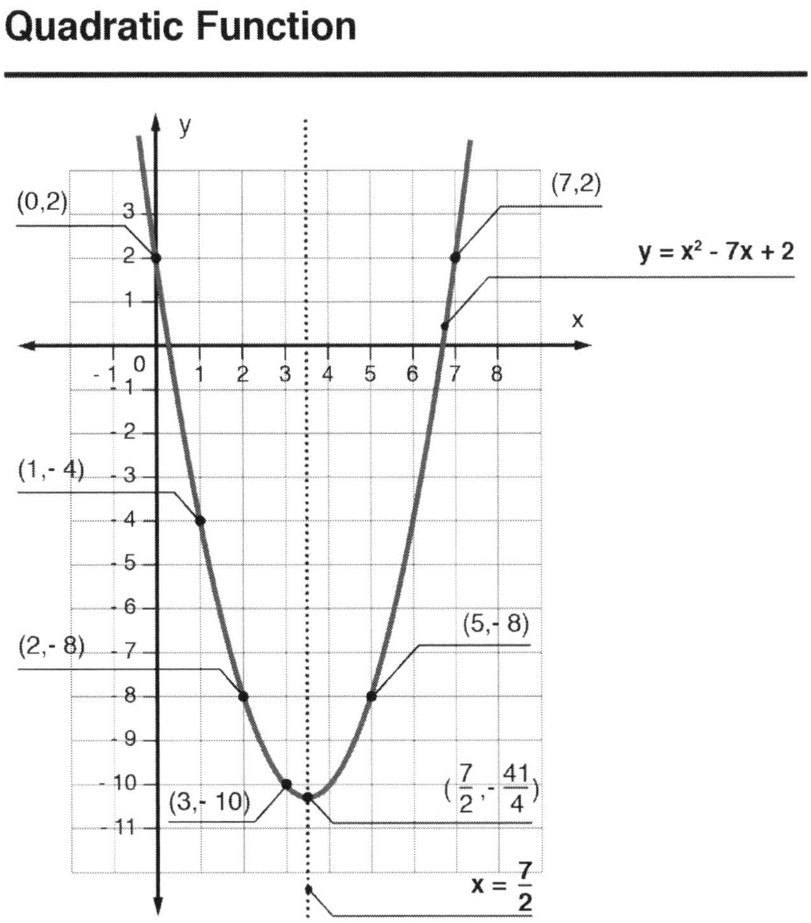


## Polynomial Functions

An expression of the form $ax^n$, where *n* is a non-negative integer, is called a **monomial** because it contains one term. A sum of monomials is called a **polynomial.** For example, $-4x^3 + x$ is a polynomial, while $5x^7$ is a monomial. A function equal to a polynomial is called a **polynomial function**.

The monomials in a polynomial are also called the **terms** of the polynomial.

The constants that precede the variables are called **coefficients.**

The highest value of the exponent of *x* in a polynomial is called the **degree** of the polynomial. So, $-4x^3 + x$ has a degree of 3, while $-2x^5 + x^3 + 4x + 1$ has a degree of 5. When multiplying polynomials, the degree of the result will be the sum of the degrees of the two polynomials being multiplied.

To add polynomials, add the coefficients of like powers of *x*. For example:

$$(-2x^5 + x^3 + 4x + 1) + (-4x^3 + x)$$

$$(-2 + 0)x^5 + (1 - 4)x^3 + (4 + 1)x + (1 + 0)$$

$$-2x^5 - 3x^3 + 5x + 1$$

Likewise, subtraction of polynomials is performed by subtracting coefficients of like powers of *x*. So,

$$(-2x^5 + x^3 + 4x + 1) - (-4x^3 + x)$$

$$(-2 - 0)x^5 + (1 + 4)x^3 + (4 - 1)x + (1 - 0)$$

$$-2x^5 + 5x^3 + 3x + 1$$

To multiply two polynomials, multiply each term of the first polynomial by each term of the second polynomial and add the results. For example:

$$(4x^2 + x)(-x^3 + x)$$

$$4x^2(-x^3) + 4x^2(x) + x(-x^3) + x(x)$$

$$-4x^5 + 4x^3 - x^4 + x^2$$

In the case where each polynomial has two terms, like in this example, some students find it helpful to remember this as multiplying the First terms, then the Outer terms, then the Inner terms, and finally the Last terms, with the mnemonic **FOIL**. For longer polynomials, the multiplication process is the same, but there will be, of course, more terms, and there is no common mnemonic to remember each combination.

The process of **factoring** a polynomial means to write the polynomial as a product of other (generally simpler) polynomials. Here is an example:

$$x^2 - 4x + 3 = (x - 1)(x - 3)$$

Factors for polynomials are similar to factors for integers—they are numbers, variables, or polynomials that, when multiplied together, give a product equal to the polynomial in question. One polynomial is a factor of a second polynomial if the second polynomial can be obtained from the first by multiplying by a third polynomial.

$6x^6 + 13x^4 + 6x^2$ can be obtained by multiplying together $(3x^4 + 2x^2)(2x^2 + 3)$. This means $2x^2 + 3$ and $3x^4 + 2x^2$ are factors of $6x^6 + 13x^4 + 6x^2$.

In general, finding the factors of a polynomial can be tricky. However, there are a few types of polynomials that can be factored in a straightforward way.

If a certain monomial is in each term of a polynomial, it can be factored out. There are several common forms polynomials take, which if you recognize, you can solve. The first example is a perfect square trinomial. To factor this polynomial, first expand the middle term of the expression:

$$x^2 + 2xy + y^2$$

$$x^2 + xy + xy + y^2$$

Factor out a common term in each half of the expression (in this case $x$ from the left and $y$ from the right):

$$x(x + y) + y(x + y)$$

Then the same can be done again, treating $(x + y)$ as the common factor:

$$(x + y)(x + y) = (x + y)^2$$

Therefore, the formula for this polynomial is:

$$x^2 + 2xy + y^2 = (x + y)^2$$

Next is another example of a perfect square trinomial. The process is the similar, but notice the difference in sign:

$$x^2 - 2xy + y^2$$

$$x^2 - xy - xy + y^2$$

Factor out the common term on each side:

$$x(x - y) - y(x - y)$$

Factoring out the common term again:

$$(x - y)(x - y) = (x - y)^2$$

Thus:

$$x^2 - 2xy + y^2 = (x - y)^2$$

The next is known as a difference of squares. This process is effectively the reverse of binomial multiplication:

$$x^2 - y^2$$

$$x^2 - xy + xy - y^2$$

$$x(x - y) + y(x - y)$$

$$(x + y)(x - y)$$

Therefore:

$$x^2 - y^2 = (x + y)(x - y)$$

The following two polynomials are known as the sum or difference of cubes. These are special polynomials that take the form of $x^3 + y^3$ or $x^3 - y^3$. The following formula factors the sum of cubes:

$$x^3 + y^3 = (x + y)(x^2 - xy + y^2)$$

Next is the difference of cubes, but note the change in sign. The formulas for both are similar, but the order of signs for factoring the sum or difference of cubes can be remembered by using the acronym SOAP, which stands for "same, opposite, always positive." The first sign is the same as the sign in the first expression, the second is opposite, and the third is always positive. The next formula factors the difference of cubes:

$$x^3 - y^3 = (x - y)(x^2 + xy + y^2)$$

The following two examples are expansions of cubed binomials. Similarly, these polynomials always follow a pattern:

$$x^3 + 3x^2y + 3xy^2 + y^3 = (x + y)^3$$

$$x^3 - 3x^2y + 3xy^2 - y^3 = (x - y)^3$$

These rules can be used in many combinations with one another. For example, the expression $3x^3 - 24$ has a common factor of 3, which becomes:

$$3(x^3 - 8)$$

A difference of cubes still remains which can then be factored out:

$$3(x - 2)(x^2 + 2x + 4)$$

There are no other terms to be pulled out, so this expression is completely factored.

When factoring polynomials, a good strategy is to multiply the factors to check the result. Let's try another example:

$$4x^3 + 16x^2$$

Both sides of the expression can be divided by 4, and both contain $x^2$, because $4x^3$ can be thought of as $4x^2(x)$, so the common term can simply be factored out:

$$4x^2(x + 4)$$

It sometimes can be necessary to rewrite the polynomial in some clever way before applying the above rules. Consider the problem of factoring $x^4 - 1$. This does not immediately look like any of the previous polynomials. However, it's possible to think of this polynomial as $x^4 - 1 = (x^2)^2 - (1^2)^2$, and now it can be treated as a difference of squares to simplify this:

$$(x^2)^2 - (1^2)^2$$

$$(x^2)^2 - x^2 1^2 + x^2 1^2 - (1^2)^2$$

$$x^2(x^2 - 1^2) + 1^2(x^2 - 1^2)$$

$$(x^2 + 1^2)(x^2 - 1^2)$$

$$(x^2 + 1)(x^2 - 1)$$

### *Zeros of Polynomials*

Finding the zeros of polynomial functions is the same process as finding the solutions of polynomial equations. These are the points at which the graph of the function crosses the *x*-axis. Factors can be used to find the zeros of a polynomial function. The degree of the function shows the number of possible zeros. If the highest exponent on the independent variable is 4, then the degree is 4, and the number of possible zeros is 4. If there are complex solutions, the number of roots is less than the degree.

Given the function $y = x^2 + 7x + 6$, $y$ can be set equal to zero, and the polynomial can be factored. The equation turns into $0 = (x + 1)(x + 6)$, where $x = -1$ and $x = -6$ are the zeros. Since this is a quadratic equation, the shape of the graph will be a parabola. Knowing that zeros represent the points where the parabola crosses the *x*-axis, the maximum or minimum point is the only other piece needed to sketch a rough graph of the function. By looking at the function in standard form, the coefficient of $x$ is positive; therefore, the parabola opens up. Using the zeros and the minimum, the following rough sketch of the graph can be constructed:

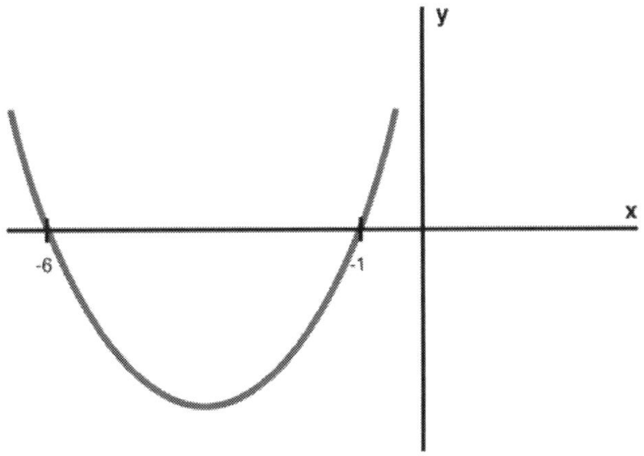


## Rational Functions

A **rational equation** can be as simple as an equation with a ratio of polynomials, $\frac{p(x)}{q(x)}$, set equal to a value, where $p(x)$ and $q(x)$ are both polynomials. Notice that a rational equation has an equal sign, which is different from expressions. This leads to solutions, or numbers that make the equation true.

It is possible to solve rational equations by trying to get all of the *x* terms out of the denominator and then isolate them on one side of the equation. For example, to solve the equation $\frac{3x+2}{2x+3} = 4$, start by multiplying both sides by $(2x + 3)$. This will cancel on the left side to yield $3x + 2 = 4(2x + 3)$, then:

$$3x + 2 = 8x + 12$$

Now, subtract $8x$ from both sides, which yields:

$$-5x + 2 = 12$$

Subtracting 2 from both sides results in $-5x = 10$. Finally, divide both sides by -5 to obtain $x = -2$.

Sometimes, when solving rational equations, it can be easier to try to simplify the rational expression by factoring the numerator and denominator first, then cancelling out common factors. For example, to solve:

$$\frac{2x^2 - 8x + 6}{x^2 - 3x + 2} = 1$$

Start by factoring:

$$2x^2 - 8x + 6 = 2(x^2 - 4x + 3) = 2(x - 1)(x - 3)$$

Then, factor $x^2 - 3x + 2$ into $(x - 1)(x - 2)$. This turns the original equation into:

$$\frac{2(x - 1)(x - 3)}{(x - 1)(x - 2)} = 1$$

The common factor of $(x - 1)$ can be canceled, leaving us with:

$$\frac{2(x - 3)}{x - 2} = 1$$

Now the same method used in the previous example can be followed. Multiplying both sides by $x - 2$ and performing the multiplication on the left yields $2x - 6 = x - 2$, which can be simplified to $x = 4$.

A **rational function** is similar to an equation, but it includes two variables. In general, a rational function is in the form: $f(x) = \frac{p(x)}{q(x)}$, where $p(x)$ and $q(x)$ are polynomials. Rational functions are defined everywhere except where the denominator is equal to zero. When the denominator is equal to zero, this indicates either a hole in the graph or an asymptote. An asymptote, which will be described in more detail in the following section, can be either vertical, horizontal, or slant. A hole occurs when both the numerator and denominator are equal to 0 for a given value of *x*. A rational function can have at most one vertical asymptote and one horizontal or slant asymptote. An asymptote is a line such that the distance between the curve and the line tends toward 0, but never reaches it, as the line heads toward infinity. Examples of these types of functions are shown below. The first graph shows a rational function with a vertical asymptote at x = 0.

This can be found by setting the denominator equal to 0. In this case it is just x = 0.

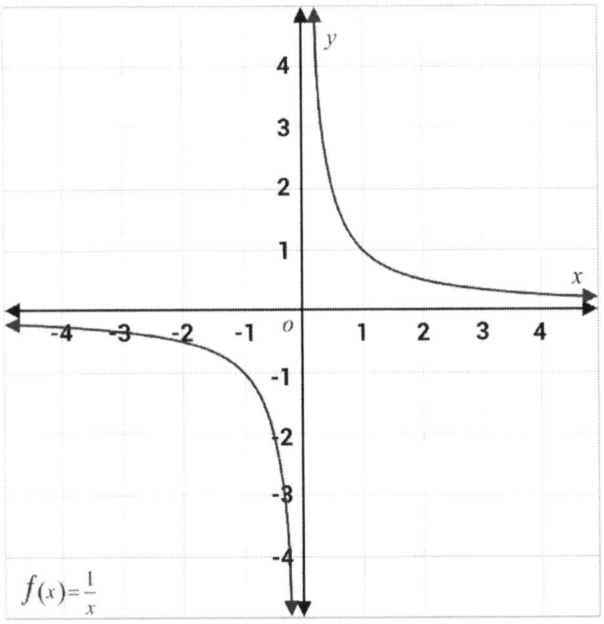


The second graph shows a rational function with a vertical asymptote at x = -.5. Again this can be found by just setting the denominator equal to 0. So, $2x^2 + x = 0, 2x + 1 = 0, 2x = -1, x = -.5$. This graph also has a hole in the graph at $x = 0$. This is because both the numerator and denominator are equal to 0 when $x = 0$.

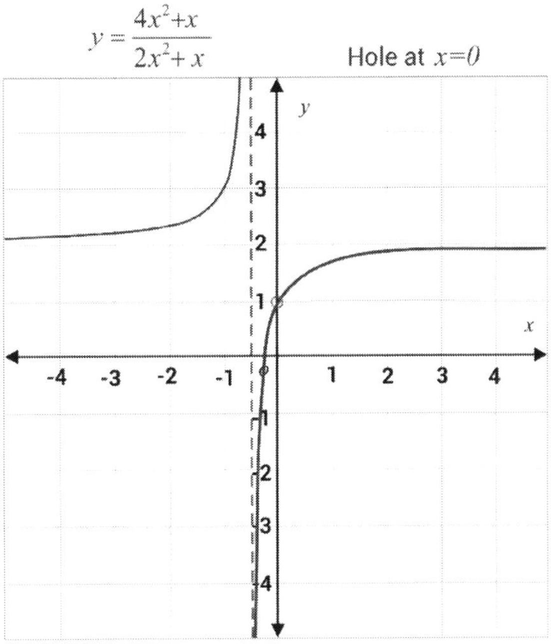

### *Asymptotes*

An **asymptote** is a line that approaches the graph of a given function, but never meets it. Vertical asymptotes correspond to denominators of zero for a rational function. They also exist in logarithmic functions and trigonometric functions, such as tangent and cotangent. In rational functions and trigonometric functions, the asymptotes exist at $x$-values that cause a denominator equal to zero. For example, vertical asymptotes exist at $x = \pm 2$ for the function:

$$f(x) = \frac{x+1}{(x-2)(x+2)}$$

Horizontal and oblique asymptotes correspond to the behavior of a curve as the $x$-values approach either positive or negative infinity. For example, the graph of $f(x) = e^x$ has a horizontal asymptote of $y = 0$ as $x$ approaches negative infinity. In regards to rational functions, there is a rule to follow. Consider the following rational function:

$$f(x) = \frac{ax^n + \cdots}{bx^m + \cdots}$$

The numerator is an nth degree polynomial and the denominator is an mth degree polynomial. If $m < n$, the line $y = 0$ is a horizontal asymptote. If $n = m$, the line $y = \frac{a}{b}$ is a horizontal asymptote. If $m > n$, then there is an oblique asymptote. In order to find the equation of the oblique asymptote, the denominator is divided into the numerator using long division. The result, minus the remainder, gives the equation of the oblique asymptote.

Here is a graph that shows an example of both a slant and a vertical asymptote:

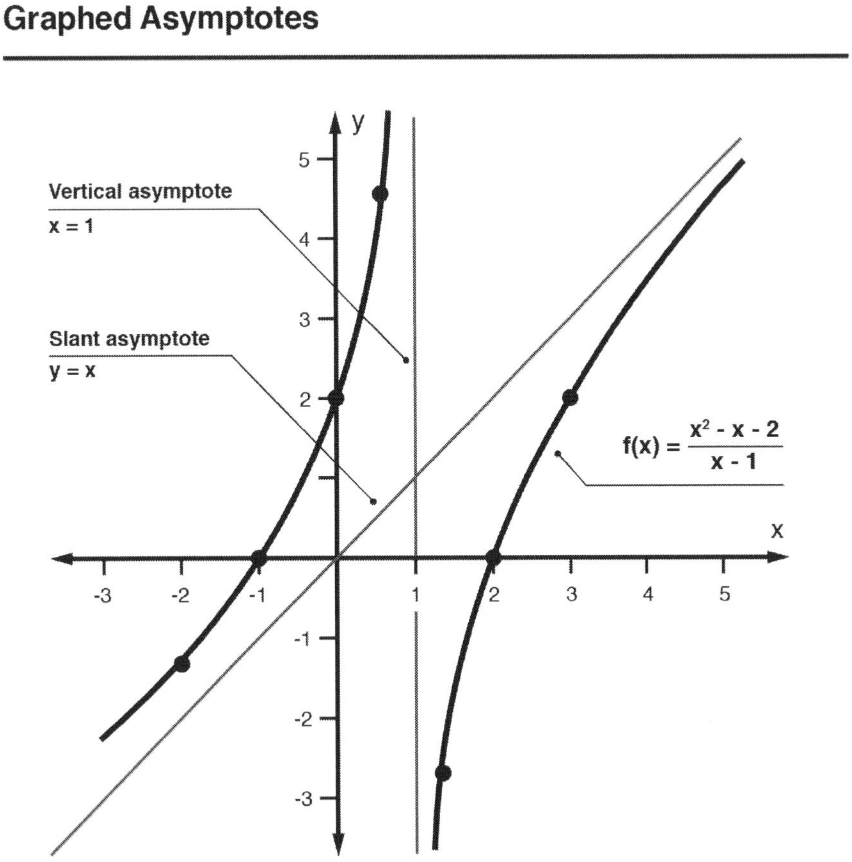

## Exponential Functions

The general form for an **exponential function** is $y = ab^x$ where *b* must be a positive number and cannot equal 1. When the value of *b* is greater than 1, the function models exponential **growth** (as *x* increases, *y* increases). When the value of *b* is less than 1, the function models exponential **decay** (as *x* increases, *y* decreases). If *a* is positive, the graph consists of points above the *x*-axis; and if *a* is negative, the graph consists of points below the *x*-axis.

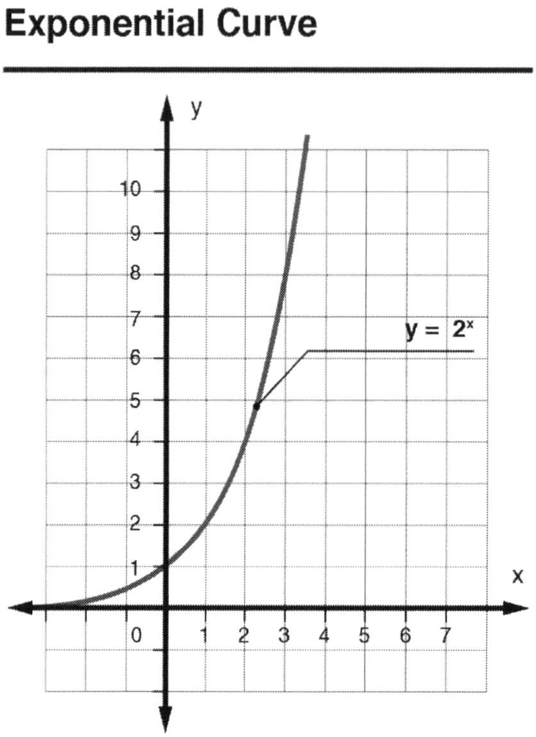


To solve an equation involving exponential expressions, the goal is to isolate the exponential expression. Once this process is completed, the logarithm—with the base equaling the base of the exponent of both sides—needs to be taken to get an expression for the variable. If the base is e, the natural log of both sides needs to be taken.

The domain of an exponential function is all real numbers, and the range is all positive real numbers. There will always be a horizontal asymptote of $y = 0$ on one side. If *b* is greater than 1, then the graph will be increasing when moving to the right. If *b* is less than 1, then the graph will be decreasing when moving to the right. Exponential functions are one-to-one. The basic exponential function graph will go through the point (0, 1).

The following example demonstrates this more clearly:

Solve $5^{x+1} = 25$.

The first step is to get the $x$ out of the exponent by rewriting the equation $5^{x+1} = 5^2$ so that both sides have a base of 5. Since the bases are the same, the exponents must be equal to each other. This leaves $x + 1 = 2$ or $x = 1$. To check the answer, the $x$-value of 1 can be substituted back into the original equation.

Exponential growth and decay can be found in real-world situations. For example, if a piece of notebook paper is folded 25 times, the thickness of the paper can be found. To model this situation, a table can be used. The initial point is one-fold, which yields a thickness of 2 papers. For the second fold, the thickness is 4. Since the thickness doubles each time, the table below shows the thickness for the next few folds. Notice the thickness changes by the same factor each time. Since this change for a constant interval of folds is a factor of 2, the function is exponential. The equation for this is $y = 2^x$. For twenty-five folds, the thickness would be 33,554,432 papers.

| $x$ **(folds)** | $y$ **(paper thickness)** |
|---|---|
| 0 | 1 |
| 1 | 2 |
| 2 | 4 |
| 3 | 8 |
| 4 | 16 |
| 5 | 32 |

When working with logarithms and exponential expressions, it is important to remember the relationship between the two. In general, the logarithmic form is $y = log_b x$ for an exponential form $b^y = x$. Logarithms and exponential functions are inverses of each other.

### *Graphing an Exponential Function*

To graph the curve for an exponential function, *x*-values are selected and then substituted into the function to obtain the corresponding *y*-values. A general rule of thumb is to select three negative values, zero, and three positive values. Plotting the seven points on the graph for an exponential function should allow for constructing a smooth curve through them.

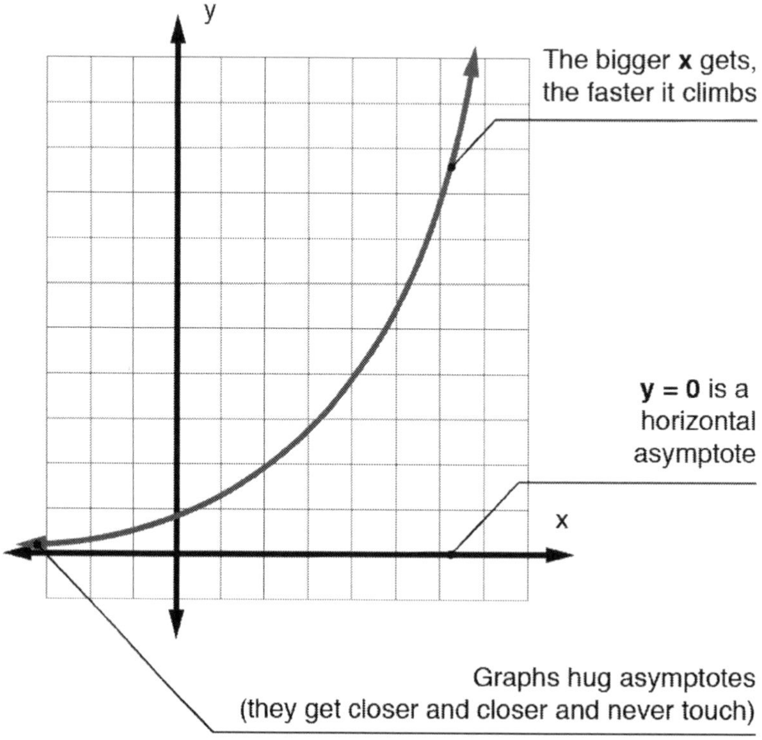


## Logarithmic Functions

A **logarithmic function** is an inverse for an exponential function. The inverse of the base *b* exponential function is written as $\log_b(x)$, and is called the **base b logarithm**. The domain of a logarithm is all positive real numbers. It has the properties that:

$$\log_b(b^x) = x$$

For positive real values of *x*:

$$b^{\log_b(x)} = x$$

When there is no chance of confusion, the parentheses are sometimes skipped for logarithmic functions: $\log_b(x)$ may be written as $\log_b x$. For the special number *e*, the base *e* logarithm is called the **natural logarithm** and is written as $\ln x$. Logarithms are one-to-one.

When working with logarithmic functions, it is important to remember the following properties. Each one can be derived from the definition of the logarithm as the inverse to an exponential function:

$$\log_b 1 = 0$$
$$\log_b b = 1$$
$$\log_b b^p = p$$
$$\log_b MN = \log_b M + \log_b N$$
$$\log_b \frac{M}{N} = \log_b M - \log_b N$$
$$\log_b M^p = p \log_b M$$

When solving equations involving exponentials and logarithms, the following fact should be used:

If $f$ is a one-to-one function, $a = b$ is equivalent to $f(a) = f(b)$.

Using this, together with the fact that logarithms and exponentials are inverses, allows for manipulations of the equations to isolate the variable as is demonstrated in the following example:

Solve $4 = \ln(x - 4)$.

Using the definition of a logarithm, the equation can be changed to:
$e^4 = e^{\ln(x-4)}$. The functions on the right side cancel with a result of $e^4 = x - 4$. This then gives $x = 4 + e^4$.

## Trigonometric Functions

Trigonometric functions are also used to describe behavior in mathematics. **Trigonometry** is the relationship between the angles and sides of a triangle. **Trigonometric functions** include sine, cosine, tangent, secant, cosecant, and cotangent. The functions are defined through ratios in a right triangle. **SOHCAHTOA** is a common acronym used to remember these ratios, which are defined by the relationships of the sides and angles relative to the right angle. **Sine** is opposite over hypotenuse, **cosine** is adjacent over hypotenuse, and **tangent** is opposite over adjacent. These ratios are the reciprocals of secant, cosecant, and cotangent, respectively. Angles can be measured in degrees or radians. Here is a diagram of SOHCAHTOA:

SOH $\sin\theta = \frac{\text{opposite}}{\text{hypotenuse}}$

CAH $\cos\theta = \frac{\text{adjacent}}{\text{hypotenuse}}$

TOA $\tan\theta = \frac{\text{opposite}}{\text{adjacent}}$

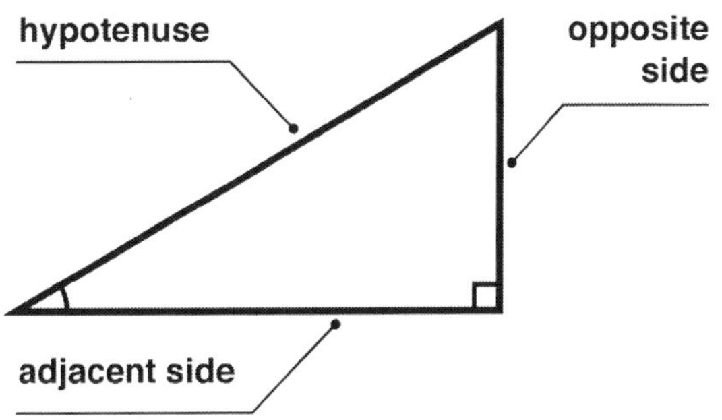


A **radian** is equal to the angle that subtends the arc with the same length as the radius of the circle. It is another unit for measuring angles, in addition to degrees. The **unit circle** is used to describe different radian measurements and the trigonometric ratios for special angles. The circle has a center at the origin, $(0, 0)$, and a radius of 1, which can be seen below. The points where the circle crosses an axis are labeled.

The circle begins on the right-hand side of the *x*-axis at 0 radians. Since the circumference of a circle is $2\pi r$ and the radius $r = 1$, the circumference is $2\pi$. Zero and $2\pi$ are labeled as radian measurements at the point $(1, 0)$ on the graph. The radian measures around the rest of the circle are labeled also in relation to $\pi$; $\pi$ is at the point $(-1, 0)$, also known as 180 degrees. Since these two measurements are equal, $\pi = 180$ degrees written as a ratio can be used to convert degrees to radians or vice versa. For example, to convert 30 degrees to radians, $30 \text{ degrees} \times \frac{\pi}{180 \text{ degrees}}$ can be used to obtain $\frac{1}{6}\pi$ or $\frac{\pi}{6}$. This radian measure is a point the unit circle

The coordinates labeled on the unit circle are found based on two common right triangles. The ratios formed in the coordinates can be found using these triangles. Each of these triangles can be inserted into the circle to correspond 30, 45, and 60 degrees or $\frac{\pi}{6}, \frac{\pi}{4}$, and $\frac{\pi}{3}$ radians.

By letting the hypotenuse length of these triangles equal 1, these triangles can be placed inside the unit circle. These coordinates can be used to find the trigonometric ratio for any of the radian measurements on the circle.

## The Unit Circle

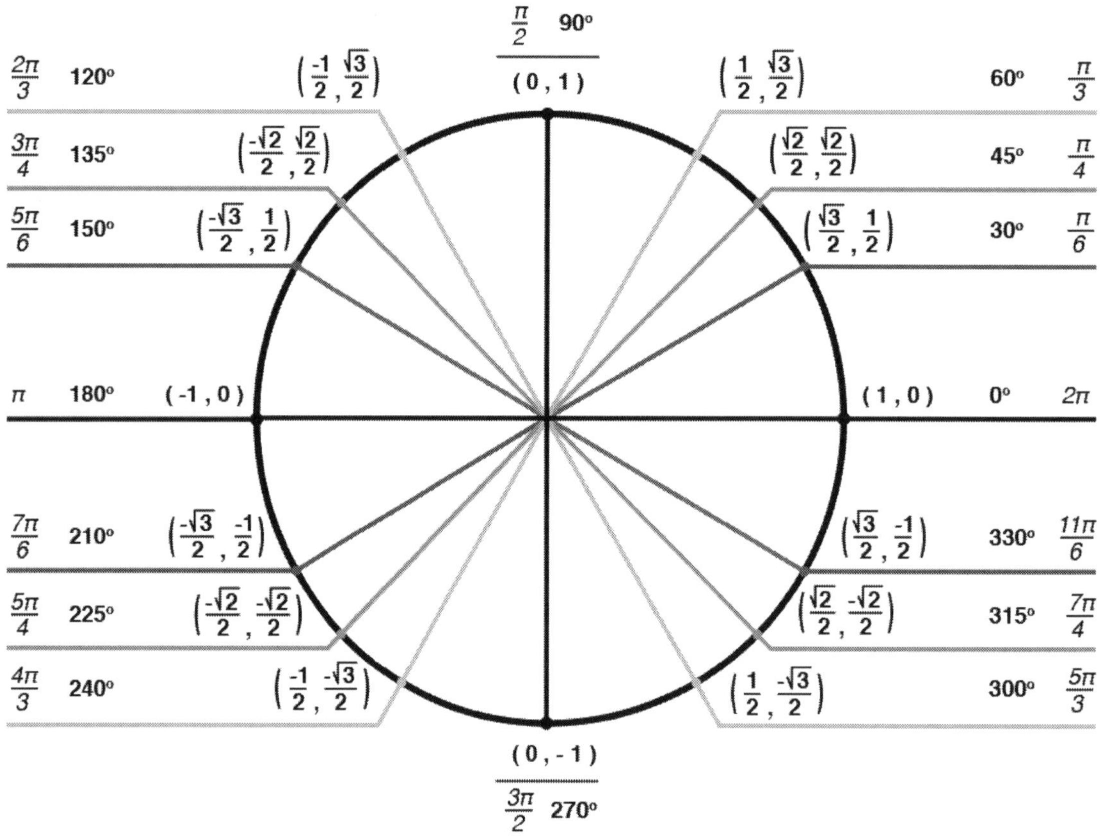

Given any $(x, y)$ on the unit circle, $\sin(\theta) = y$, $\cos(\theta) = x$, and $\tan(\theta) = \frac{y}{x}$. The value $\theta$ is the angle that spans the arc around the unit circle. For example, finding $\sin(\frac{\pi}{4})$ means finding the y-value corresponding to the angle $\theta = \frac{\pi}{4}$. The answer is $\frac{\sqrt{2}}{2}$. Finding $\cos(\frac{\pi}{3})$ means finding the $x$-value corresponding to the angle $\theta = \frac{\pi}{3}$. The answer is $\frac{1}{2}$ or 0.5. Both angles lie in the first quadrant of the unit circle. Trigonometric ratios can also be calculated for radian measures past $\frac{\pi}{2}$, or 90 degrees. Since the same special angles can be moved around the circle, the results only differ with a change in sign. This can be seen at two points labeled in the second and third quadrant.

From the unit circle definition, the sine function starts at 0 when $\theta = 0$. It grows to 1 as $\theta$ grows to $\pi/2$, and then back to 0 at $\theta = \pi$. Then it decreases to -1 as $\theta$ grows to $3\pi/2$, and goes back up to 0 at $\theta = 2\pi$.

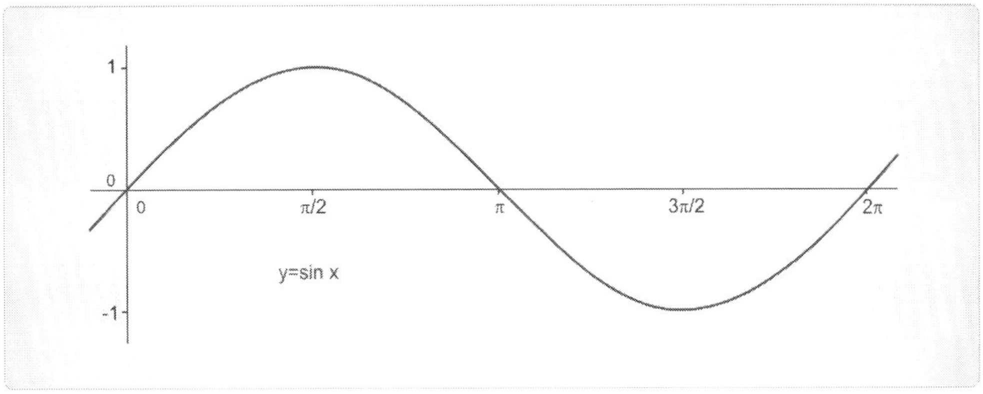


The graph of the cosine is similar. The cosine graph will start at 1, decreasing to 0 at $\pi/2$ and continuing to decrease to -1 at $\theta = \pi$. Then, it grows to 0 as $\theta$ grows to $3\pi/2$ and back up to 1 at $\theta = 2\pi$.

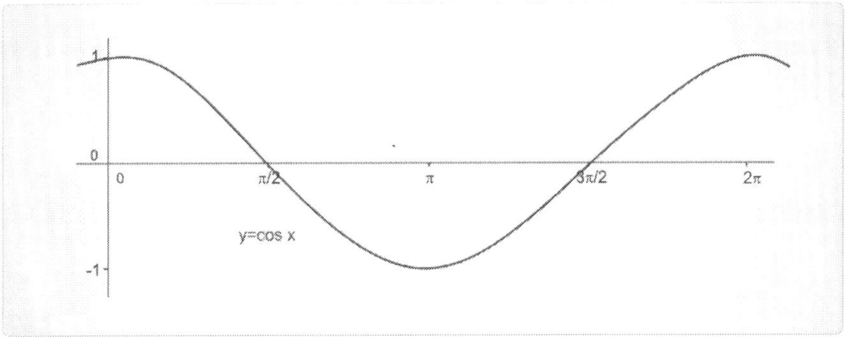


Another trigonometric function that is frequently used, is the **tangent** function. This is defined as the following equation:

$$\tan\theta = \frac{\sin\theta}{\cos\theta}$$

The tangent function is a period of $\pi$ rather than $2\pi$ because the sine and cosine functions have the same absolute values after a change in the angle of $\pi$, but they flip their signs. Since the tangent is a ratio of the two functions, the changes in signs cancel.

The tangent function will be zero when sine is zero, and it will have a vertical asymptote whenever cosine is zero. The following graph shows the tangent function:

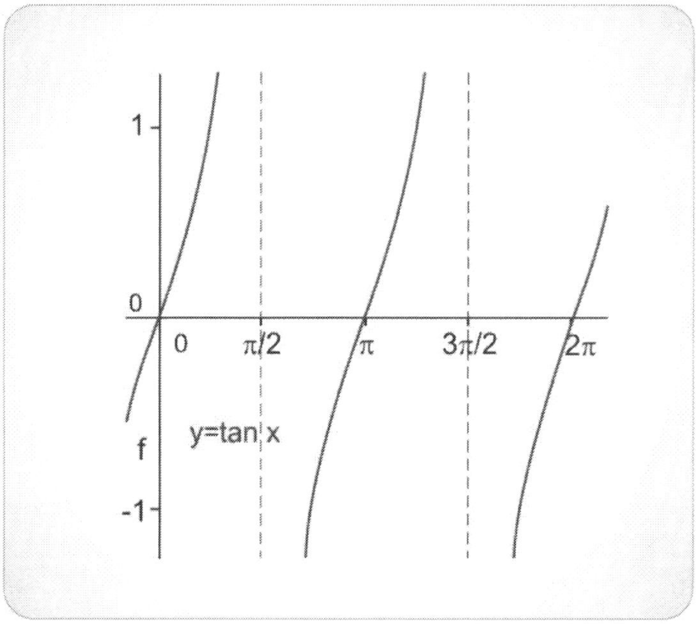


Three other trigonometric functions are sometimes useful. These are the **reciprocal** trigonometric functions, so named because they are just the reciprocals of sine, cosine, and tangent. They are the **cosecant**, defined as $\csc\theta = \frac{1}{\sin\theta}$, the **secant**, $\sec\theta = \frac{1}{\cos\theta}$, and the **cotangent**, $\cot\theta = \frac{1}{\tan\theta}$. Note that from the definition of tangent, $\cot\theta = \frac{\cos\theta}{\sin\theta}$.

In addition, there are three identities that relate the trigonometric functions to one another:

$$\cos\theta = \sin(\frac{\pi}{2} - \theta)$$

$$\csc\theta = \sec\left(\frac{\pi}{2} - \theta\right)$$

$$\cot\theta = \tan(\frac{\pi}{2} - \theta)$$

Here is a list of commonly-needed values for trigonometric functions, given in radians, for the first quadrant:

## Table for trigonometric functions

| | | |
|---|---|---|
| $\sin 0 = 0$ | $\cos 0 = 1$ | $\tan 0 = 0$ |
| $\sin\frac{\pi}{6} = \frac{1}{2}$ | $\cos\frac{\pi}{6} = \frac{\sqrt{3}}{2}$ | $\tan\frac{\pi}{6} = \frac{\sqrt{3}}{3}$ |
| $\sin\frac{\pi}{4} = \frac{\sqrt{2}}{2}$ | $\cos\frac{\pi}{4} = \frac{\sqrt{2}}{2}$ | $\tan\frac{\pi}{4} = 1$ |
| $\sin\frac{\pi}{3} = \frac{\sqrt{3}}{2}$ | $\cos\frac{\pi}{3} = \frac{1}{2}$ | $\tan\frac{\pi}{3} = \sqrt{3}$ |
| $\sin\frac{\pi}{2} = 1$ | $\cos\frac{\pi}{2} = 0$ | $\tan\frac{\pi}{2} = undefined$ |
| $\csc 0 = undefined$ | $\sec 0 = 1$ | $\cot 0 = undefined$ |
| $\csc\frac{\pi}{6} = 2$ | $\sec\frac{\pi}{6} = \frac{2\sqrt{3}}{3}$ | $\cot\frac{\pi}{6} = \sqrt{3}$ |
| $\csc\frac{\pi}{4} = \sqrt{2}$ | $\sec\frac{\pi}{4} = \sqrt{2}$ | $\cot\frac{\pi}{4} = 1$ |
| $\csc\frac{\pi}{3} = \frac{2\sqrt{3}}{3}$ | $\sec\frac{\pi}{3} = 2$ | $\cot\frac{\pi}{3} = \frac{\sqrt{3}}{3}$ |
| $\csc\frac{\pi}{2} = 1$ | $\sec\frac{\pi}{2} = undefined$ | $\cot\frac{\pi}{2} = 0$ |

To find the trigonometric values in other quadrants, complementary angles can be used. The **complementary angle** is the smallest angle between the $x$-axis and the given angle.

Once the complementary angle is known, the following rule is used:

> For an angle $\theta$ with complementary angle $x$, the absolute value of a trigonometric function evaluated at $\theta$ is the same as the absolute value when evaluated at $x$.

The correct sign for sine and cosine is determined by the $x$- and $y$-coordinates on the unit circle.

- Sine will be positive in quadrants I and II and negative in quadrants III and IV.
- Cosine will be positive in quadrants I and IV, and negative in quadrants II and III.
- Tangent will be positive in I and III, and negative in quadrants II and IV.
- The signs of the reciprocal functions will be the same as the sign of the function for which they are the reciprocal.

For example: Find $\sin\frac{3\pi}{4}$.

> The reference angle must be found first. This angle is in the II quadrant, and the angle between it and the $x$-axis is $\frac{\pi}{4}$. Now, $\sin\frac{\pi}{4} = \frac{\sqrt{2}}{2}$. Since this is in the II quadrant, sine takes on positive values (the $y$-coordinate is positive in quadrant II). Therefore, $\sin\frac{3\pi}{4} = \frac{\sqrt{2}}{2}$.

When solving equations that involve trigonometric functions, there are often multiple solutions. For example, $2\sin\theta = \sqrt{2}$ can be simplified to $\sin\theta = \frac{\sqrt{2}}{2}$. This has solutions $\theta = \frac{\pi}{4}, \frac{3\pi}{4}$, but in addition, because of the periodicity, any integer multiple of $2\pi$ can also be added to these solutions to find another solution.

The full set of solutions is $\theta = \frac{\pi}{4} + 2\pi k, \frac{3\pi}{4} + 2\pi k$ for all integer values of $k$. It is very important to remember to find all possible solutions when dealing with equations that involve trigonometric functions.

The name *trigonometric* comes from the fact that these functions play an important role in the geometry of triangles, particularly right triangles. Consider the right triangle shown in this figure:

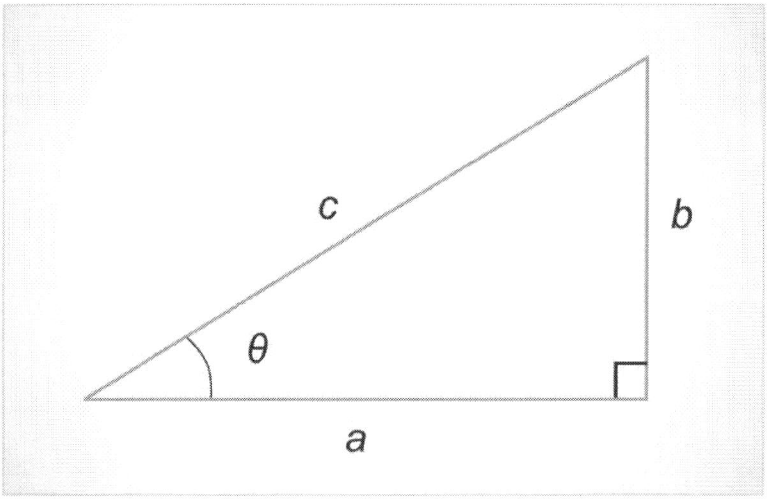


The following hold true:

- $c\sin\theta = b$.
- $c\cos\theta = a$.
- $\tan\theta = \frac{b}{a}$.
- $b\csc\theta = c$.
- $a\sec\theta = c$.
- $\cot\theta = \frac{a}{b}$.

It is important to remember that the angles of a triangle must add up to $\pi$ radians (180 degrees).

The graph of trigonometric functions can be used to model different situations. General forms are

$$y = a \sin b(x - h) + k$$

and

$$y = a \cos b\,(x - h) + k$$

The variable $a$ represents the **amplitude**, which shows the maximum and minimum value of the function. The $b$ is used to find the **period** by using the ratio $\frac{2\pi}{b}$, $h$ is the **horizontal shift**, and $k$ is the **vertical shift**.

The equation $y = \sin(x)$ is shown on the following graph with the thick black line. The stretched graph of $y = 2\sin(x)$ is shown in solid black, and the shrunken graph $y = \frac{1}{2}\sin(x)$ is shown with the dotted line.

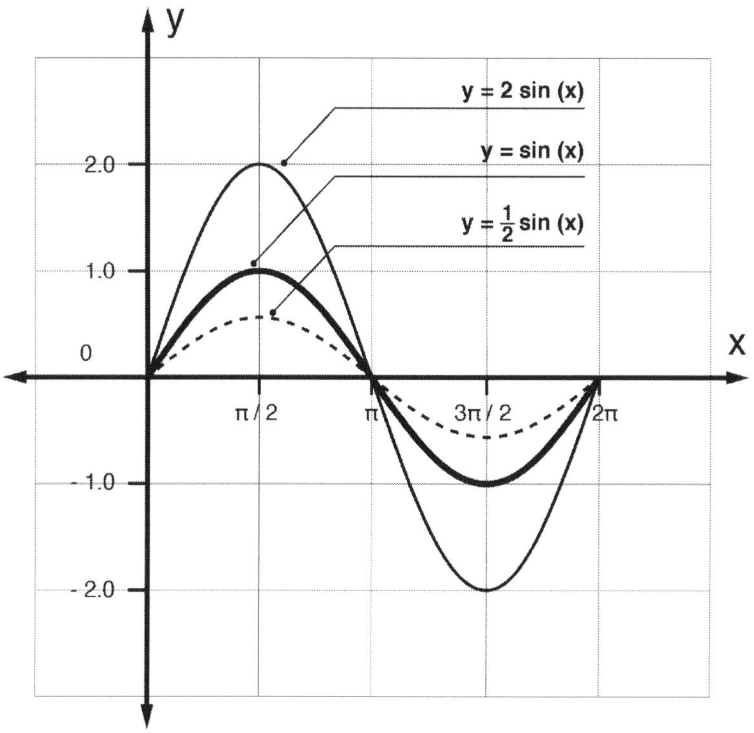


## *Solving Trigonometric Functions*

Solving trigonometric functions can be done with a knowledge of the unit circle and the trigonometric identities. It requires the use of opposite operations combined with trigonometric ratios for special triangles. For example, the problem may require solving the equation $2\cos^2 x - \sqrt{3}\cos x = 0$ for the values of $x$ between 0 and 180 degrees. The first step is to factor out the $\cos x$ term, resulting in:

$$\cos x\,(2\cos x - \sqrt{3}) = 0$$

By the factoring method of solving, each factor can be set equal to zero:

$$\cos x = 0$$

$$(2\cos x - \sqrt{3}) = 0$$

The second equation can be solved to yield the following equation:

$$\cos x = \frac{\sqrt{3}}{2}$$

Now that the value of $x$ is found, the trigonometric ratios can be used to find the solutions of:

$$x = 30 \text{ and } 90 \text{ degrees}$$

Solving trigonometric functions requires the use of algebra to isolate the variable and a knowledge of trigonometric ratios to find the value of the variable. The unit circle can be used to find answers for special triangles. Beyond those triangles, a calculator can be used to solve for variables within the trigonometric functions.

### *Representations of Trigonometric Functions*

Besides their function form, trigonometric and inverse trigonometric functions can be represented in other various formats. Regardless of the form in which they are given—such as in a table, list of numbers, or graph—it is important to note their periodic tendencies. They all return to the same function values over regular intervals in the domain. Other important concepts to notice in the various formats are range and asymptotes. Many trigonometric functions have a minimum and maximum value, but some have vertical asymptotes at regular intervals, in which the function blows up. All of these properties can be seen in any representation of such functions.

### *Using Graphing Calculators to Solve Trigonometric Problems*

In addition to algebraic techniques, problems involving trigonometric functions can be solved using graphing calculators. For example, given an equation, both sides of the equals sign first need to be graphed as separate equations in the same window on the calculator. The point(s) of intersection are then found by zooming into an appropriate window. The points of intersection are the solutions. For example, consider $\cos x = \frac{1}{2}$. Solving this equation involves graphing both $y = \cos x$ and $y = \frac{1}{2}$ in the same window. If the calculator is set in Radians mode, the screen zoomed in from $[0, 2\pi]$ on the $x$-axis would look like:

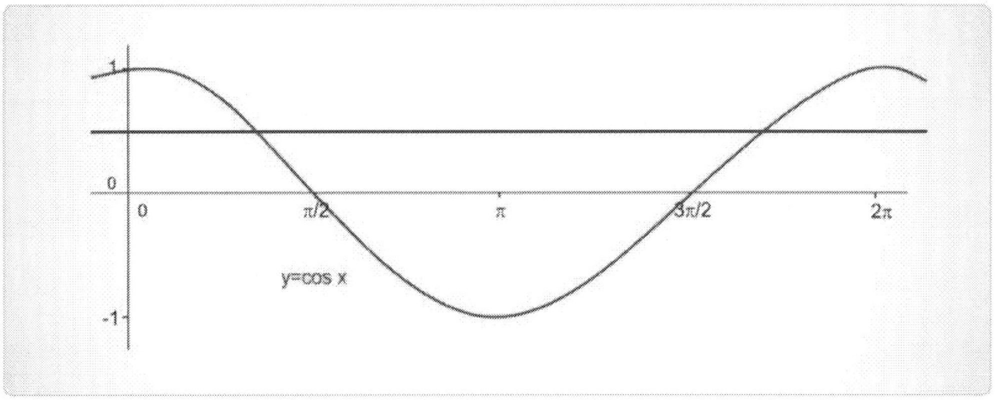

The trace function on the calculator allows the user to zoom into the point of intersection to obtain the two solutions from $[0, 2\pi]$. The periodic nature of the function must then be taken into consideration to obtain the entire solution set, which contains an infinite number of solutions.

## Inverse Trigonometric Functions

In addition to the six trigonometric functions defined above, there are inverses for these functions. The inverse functions for the trigonometric functions are also called **arc functions.** In addition to being written with a -1 as the exponent to denote that the function is an inverse, they will sometimes be written with an "a" or "arc" in front of the function name, so $\cos^{-1}\theta = \text{acos}\,\theta = \arccos\theta$.

Trigonometric functions are used to find unknown ratios for a given angle measure. The inverse of trigonometric functions is used to find the unknown angle, given a ratio. For example, the expression $\arcsin(\frac{1}{2})$ means finding the value of $x$ for $\sin(x) = \frac{1}{2}$. Since $\sin(\theta) = \frac{y}{1}$ on the unit circle, the angle whose $y$-value is $\frac{1}{2}$ is $\frac{\pi}{6}$. The inverse of any of the trigonometric functions can be used to find a missing angle measurement. Values not found on the unit circle can be found using the trigonometric functions on the calculator, making sure its mode is set to degrees or radians.

In order for the inverse to exist, the function must be one-to-one over its domain. There cannot be two input values connected to the same output. For example, the following graphs show the functions $y = \cos(x)$ and $y = \arccos(x)$. In order to have an inverse, the domain of cosine is restricted from 0 to $\pi$.

Therefore, the range of its inverse function is $[0, \pi]$.

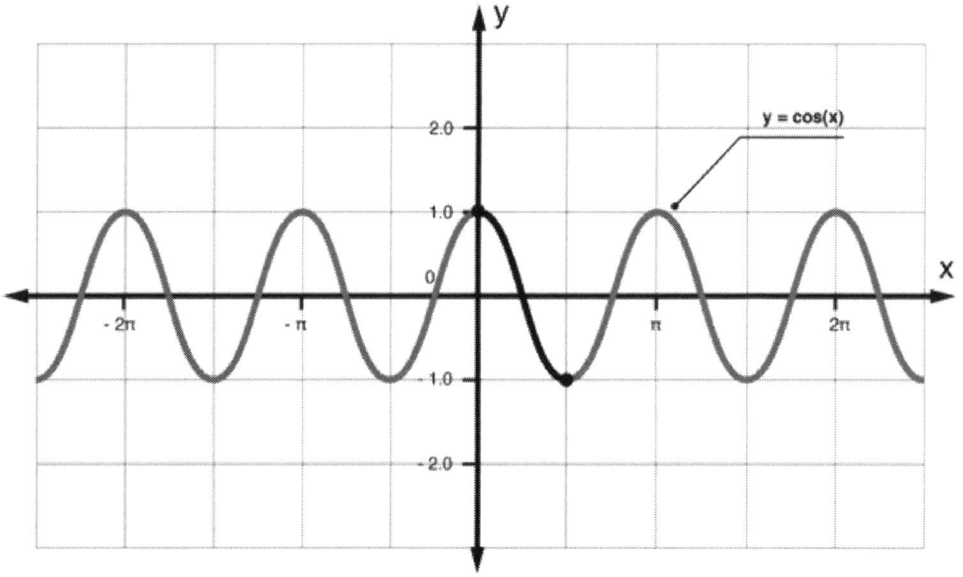


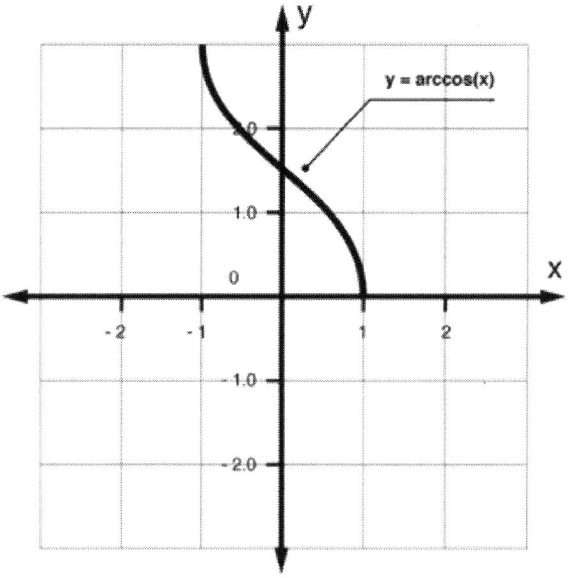


Inverses of trigonometric functions can be used to solve real-world problems. For example, there are many situations where the lengths of a perceived triangle can be found, but the angles are unknown. Consider a problem where the height of a flag (25 feet) and the distance on the ground to the flag is given (42 feet). The unknown, $x$, is the angle. To find this angle, the equation $\tan x = \frac{42}{25}$ is used. To solve for $x$, the inverse function can be used to turn the equation into:

$$\tan^{-1}\frac{42}{25} = x$$

Using the calculator, in degree mode, the answer is found to be $x = 59.2$ degrees.

## Periodic Functions

A function can be described as **periodic** if it repeats itself in regular intervals. Common periodic functions are trigonometric functions. For example, $y = \sin x$ is a periodic function with period $2\pi$ because it repeats itself every $2\pi$ units along the $x$-axis. Essentially, since the angle returns to the start every $2\pi$ radians (or 360 degrees), the graph of these functions is periodic, with period $2\pi$. This means that the graph repeats itself as one moves along the $x$-axis because:

$$\sin\theta = \sin(\theta + 2\pi)$$

Cosine works similarly. For each input angle value, the output value follows around the unit circle. Once it reaches the starting point, it continues around and around the circle. It is true that:

$$\sin(0) = \sin(2\pi) = \sin(4\pi), \text{etc.}$$

and

$$\cos(0) = \cos(2\pi) = \cos(4\pi)$$

Tangent has period $\pi$, and its output values repeat themselves every half of the unit circle. The domain of sine and cosine are all real numbers, and the domain of tangent is all real numbers, except the points where cosine equals zero. It is also true that

$$\sin(-x) = -\sin x$$

$$\cos(-x) = \cos(x)$$

$$\tan(-x) = -\tan(\mathrm{x})$$

Thus, sine and tangent are odd functions, while cosine is an even function. Sine and tangent are symmetric with respect the origin, and cosine is symmetric with respect to the *y*-axis.

## Piecewise Functions

A piecewise-defined function has a different appearance on the graph. In the following function, there are three equations defined over different intervals. It is a function because there is only one y-value for each $x$-value, passing the Vertical Line Test. The domain is all real numbers less than or equal to 6. The range is all real numbers greater than zero. From left to right, the graph decreases to zero, then increases to almost 4, and then jumps to 6. From input values greater than 2, the input decreases just below 8 to 4, and then stops.

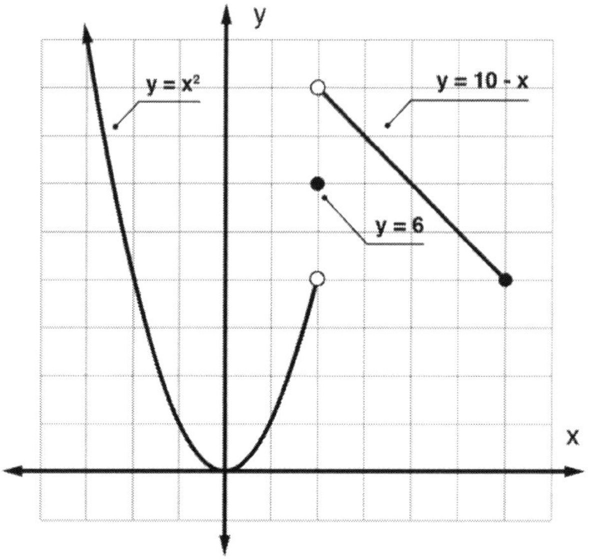


## Recursive Functions

Functions can also be defined recursively. In this form, they are not defined explicitly in terms of variables. Instead, they are defined using previously-evaluated function outputs, starting with either $f(0)$ or $f(1)$. An example of a recursively-defined function is:

$$f(1) = 2$$

$$f(n) = 2f(n-1) + 2n$$

$$n > 1$$

The domain of this function is the set of all integers.

## Parametric Functions

Parametric equations are a useful way to express two variables, such as *x* and *y*, in terms of a third variable, such as *t*. The *t* variable is known as the parameter, and *x* and *y* are functions of *t*, i.e., $x = f(t)$ and $y = g(t)$. A simple type of a parametric equation represents a line. Such an example is $x = 4 + t$ and $y = 2 - t$. Setting *t* equal to a constant results in a value for both *x* and *y*. For instance, if $t = 1$, then $x = 5$ and $y = 1$. Each equation can be solved for *t* to find the equation of the line in terms of *x* and *y* only. This process results in the two equations $t = x - 4$ and $t = y - 2$. Setting these equations equal to one another results in the equation $x - 4 = y - 2$, which, in slope-intercept form, is $y = x - 2$. This equation has slope 1 and *y*–intercept -2 because it is of the form $y = mx + b$.

Curves can also be represented by parametric equations. For instance, consider the two equations $x = t - 2$ and $y = 2t^2$. The corresponding equation in terms of only *x* and *y*, known as the **rectangular**

**equation**, can be found by solving for $t$ in the first equation and plugging it into the second equation. The first equation has equivalent form $t = x + 2$. Plugging the right-hand side in for $t$ in the second equation gives $y = 2(x + 2)^2 = 2(x^2 + 4x + 4) = 4x^2 + 8x + 8$, which is a quadratic equation. Its graph is a parabola. Once the rectangular form of the parametric equation is found, the equation can be graphed normally in a Cartesian plane.

## Building a Function

Functions can be built out of the context of a situation. For example, the relationship between the money paid for a gym membership and the months that someone has been a member can be described through a function. If the one-time membership fee is $40 and the monthly fee is $30, then the function can be written $f(x) = 30x + 40$. The $x$-value represents the number of months the person has been part of the gym, while the output is the total money paid for the membership. The table below shows this relationship. It is a representation of the function because the initial cost is $40 and the cost increases each month by $30.

| x (months) | y (money paid to gym) |
|---|---|
| 0 | 40 |
| 1 | 70 |
| 2 | 100 |
| 3 | 130 |

Functions can also be built from existing functions. For example, a given function $f(x)$ can be transformed by adding a constant, multiplying by a constant, or changing the input value by a constant. The new function $g(x) = f(x) + k$ represents a vertical shift of the original function. In $f(x) = 3x - 2$, a vertical shift 4 units up would be:

$$g(x) = 3x - 2 + 4 = 3x + 2$$

Multiplying the function times a constant $k$ represents a vertical stretch, based on whether the constant is greater than or less than 1. The function

$$g(x) = kf(x) = 4(3x - 2) = 12x - 8$$

represents a stretch. Changing the input $x$ by a constant forms the function:

$$g(x) = f(x + k) = 3(x + 4) - 2$$

$$3x + 12 - 2 = 3x + 10$$

and this represents a horizontal shift to the left 4 units. If $(x - 4)$ was plugged into the function, it would represent a vertical shift.

A composition function can also be formed by plugging one function into another. In function notation, this is written:

$$(f \circ g)(x) = f(g(x))$$

For two functions $f(x) = x^2$ and $g(x) = x - 3$, the composition function becomes:

$$f(g(x)) = (x - 3)^2$$

$$x^2 - 6x + 9$$

The composition of functions can also be used to verify if two functions are inverses of each other. Given the two functions $f(x) = 2x + 5$ and $g(x) = \frac{x-5}{2}$, the composition function can be found $(f°g)(x)$. Solving this equation yields:

$$f(g(x)) = 2\left(\frac{x-5}{2}\right) + 5$$

$$x - 5 + 5 = x$$

It also is true that $g(f(x)) = x$. Since the composition of these two functions gives a simplified answer of $x$, this verifies that $f(x)$ and $g(x)$ are inverse functions. The domain of $f(g(x))$ is the set of all x-values in the domain of $g(x)$ such that $g(x)$ is in the domain of $f(x)$. Basically, both $f(g(x))$ and $g(x)$ have to be defined.

To build an inverse of a function, $f(x)$ needs to be replaced with $y$, and the x and y values need to be switched. Then, the equation can be solved for y. For example, given the equation $y = e^{2x}$, the inverse can be found by rewriting the equation $x = e^{2y}$. The natural logarithm of both sides is taken down, and the exponent is brought down to form the equation:

$$\ln(x) = \ln(e)\, 2y$$

$\ln(e)$=1, which yields the equation $\ln(x) = 2y$. Dividing both sides by 2 yields the inverse equation

$$\frac{\ln(x)}{2} = y = f^{-1}(x)$$

The domain of an inverse function is the range of the original function, and the range of an inverse function is the domain of the original function. Therefore, an ordered pair $(x, y)$ on either a graph or a table corresponding to $f(x)$ means that the ordered pair $(y, x)$ exists on the graph of $f^{-1}(x)$. Basically, if $f(x) = y$, then $f^{-1}(y) = x$. For a function to have an inverse, it must be one-to-one. That means it must pass the **Horizontal Line Test**, and if any horizontal line passes through the graph of the function twice, a function is not one-to-one. The domain of a function that is not one-to-one can be restricted to an interval in which the function is one-to-one, to be able to define an inverse function.

Functions can also be formed from combinations of existing functions.

Given $f(x)$ and $g(x)$, the following can be built:

$$f + g$$

$$f - g$$

$$fg$$

$$\frac{f}{g}$$

The domains of $f + g, f - g,$ and $fg$ are the intersection of the domains of $f$ and $g$. The domain of $\frac{f}{g}$ is the same set, excluding those values that make $g(x) = 0$.

For example, if:

$$f(x) = 2x + 3$$

$$g(x) = x + 1$$

then

$$\frac{f}{g} = \frac{2x + 3}{x + 1}$$

Its domain is all real numbers except -1.

## Comparing Functions

As mentioned, three common functions used to model different relationships between quantities are linear, quadratic, and exponential functions. **Linear functions** are the simplest of the three, and the independent variable $x$ has an exponent of 1. Written in the most common form, $y = mx + b$, the coefficient of $x$ tells how fast the function grows at a constant rate, and the $b$-value tells the starting point. A **quadratic** function has an exponent of 2 on the independent variable $x$.

Standard form for this type of function is $y = ax^2 + bx + c$, and the graph is a parabola. These type functions grow at a changing rate. An **exponential** function has an independent variable in the exponent $y = ab^x$. The graph of these types of functions is described as **growth** or **decay**, based on whether the base, $b$, is greater than or less than 1. These functions are different from quadratic functions because the base stays constant. A common base is base *e*.

The following three functions model a linear, quadratic, and exponential function respectively: $y = 2x$, $y = x^2$, and $y = 2^x$. Their graphs are shown below. The first graph, modeling the linear function, shows that the growth is constant over each interval. With a horizontal change of 1, the vertical change is 2. It models a constant positive growth. The second graph shows the quadratic function, which is a curve that is symmetric across the y-axis. The growth is not constant, but the change is mirrored over the axis. The last graph models the exponential function, where the horizontal change of 1 yields a vertical change that increases more and more. The exponential graph gets very close to the $x$-axis, but never touches it, meaning there is an asymptote there.

The $y$-value can never be zero because the base of 2 can never be raised to an input value that yields an output of zero.

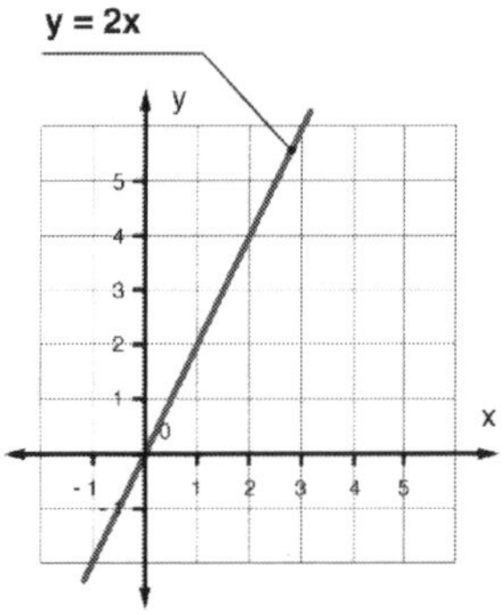


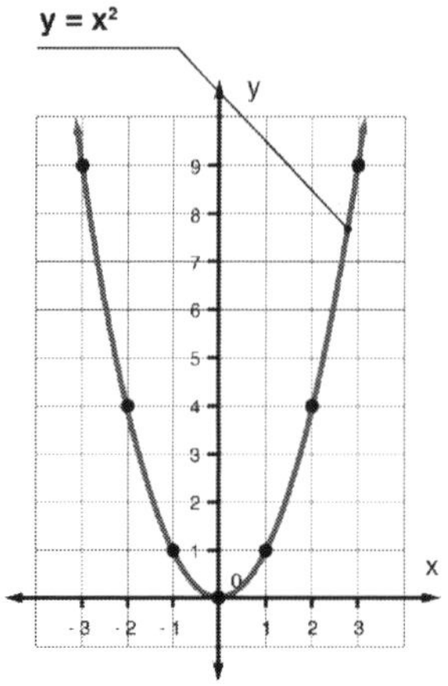


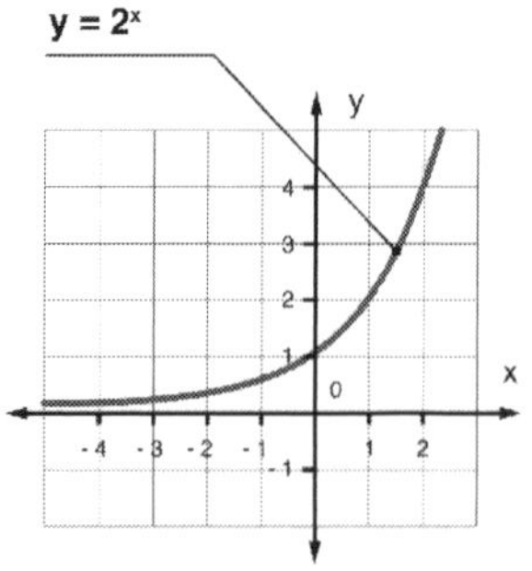


The three tables below show specific values for three types of functions. The third column in each table shows the change in the $y$-values for each interval. The first table shows a constant change of 2 for each equal interval, which matches the slope in the equation $y = 2x$. The second table shows an increasing change, but it also has a pattern. The increase is changing by 2 more each time, so the change is quadratic. The third table shows the change as factors of the base, 2. It shows a continuing pattern of factors of the base.

| $y = 2x$ | | |
|---|---|---|
| x | y | $\Delta y$ |
| 1 | 2 | |
| 2 | 4 | 2 |
| 3 | 6 | 2 |
| 4 | 8 | 2 |
| 5 | 10 | 2 |

| $y = x^2$ | | |
|---|---|---|
| x | y | $\Delta y$ |
| 1 | 1 | |
| 2 | 4 | 3 |
| 3 | 9 | 5 |
| 4 | 16 | 7 |
| 5 | 25 | 9 |

| $y = 2^x$ | | |
|---|---|---|
| x | y | $\Delta y$ |
| 1 | 2 | |
| 2 | 4 | 2 |
| 3 | 8 | 4 |
| 4 | 16 | 8 |
| 5 | 32 | 16 |

Given a table of values, the type of function can be determined by observing the change in $y$ over equal intervals. For example, the tables below model two functions. The changes in interval for the $x$-values is 1 for both tables. For the first table, the $y$-values increase by 5 for each interval. Since the change is constant, the situation can be described as a linear function. The equation would be $y = 5x + 3$. For the second table, the change for $y$ is 5, 20, 100, and 500, respectively. The increases are multiples of 5, meaning the situation can be modeled by an exponential function. The equation $y = 5^x + 3$ models this situation.

| x | y |
|---|---|
| 0 | 3 |
| 1 | 8 |
| 2 | 13 |
| 3 | 18 |
| 4 | 23 |

| x | y |
|---|---|
| 0 | 3 |
| 1 | 8 |
| 2 | 28 |
| 3 | 128 |
| 4 | 628 |

## Evaluating Functions

To evaluate functions, plug in the given value everywhere the variable appears in the expression for the function. For example, find $g(-2)$ where:

$$g(x) = 2x^2 - \frac{4}{x}$$

To complete the problem, plug in -2 in the following way:

$$g(-2) = 2(-2)^2 - \frac{4}{-2}$$

$$2 \times 4 + 2$$

$$8 + 2 = 10$$

# Practice Questions

1. What is the value of $b$ in this equation?

$$5b - 4 = 2b + 17$$

a. 13
b. 24
c. 7
d. 21
e. 9

2. If $g(x) = x^3 - 3x^2 - 2x + 6$ and $f(x) = 2$, then what is $g(f(x))$?
a. -26
b. 6
c. $2x^3 - 6x^2 - 4x + 12$
d. $-3x - 2$
e. -2

3. $(4x^2y^4)^{\frac{3}{2}}$ can be simplified to which of the following?
a. $8x^3y^6$
b. $4x^{\frac{5}{2}}y$
c. $4xy$
d. $32x^{\frac{7}{2}}y^{\frac{11}{2}}$
e. $x^3y^6$

4. What is the solution for the following equation?

$$\frac{x^2 + x - 30}{x - 5} = 11$$

a. $x = -6$
b. There is no solution.
c. $x = 16$
d. $x = 5$
e. $x = -5$

5. The phone bill is calculated each month using the equation $c = 50g + 75$. The cost of the phone bill per month is represented by $c$, and $g$ represents the gigabytes of data used that month. What is the value and interpretation of the slope of this equation?
a. 75 dollars per day
b. 75 gigabytes per day
c. 50 dollars per day
d. 50 dollars per gigabyte
e. The slope cannot be determined.

6. What is the solution for the equation $\tan(x) + 1 = 0$, where $0 \leq x < 2\pi$?

a. $x = \frac{3\pi}{4}, \frac{5\pi}{4}$

b. $x = \frac{3\pi}{4}, \frac{\pi}{4}$

c. $x = \frac{5\pi}{4}, \frac{7\pi}{4}$

d. $x = \frac{3\pi}{4}, \frac{7\pi}{4}$

e. $x = \frac{5\pi}{4}, \frac{\pi}{4}$

7. How could the following function be rewritten to identify the zeros?

$$y = 3x^3 + 3x^2 - 18x$$

a. $y = 3x(x + 3)(x - 2)$
b. $y = x(x - 2)(x + 3)$
c. $y = 3x(x - 3)(x + 2)$
d. $y = (x + 3)(x - 2)$
e. $y = 3x(x + 3)(x + 2)$

8. A ball is thrown from the top of a high hill, so that the height of the ball as a function of time is $h(t) = -16t^2 + 4t + 6$, in feet. What is the maximum height of the ball in feet?

a. 6
b. 6.25
c. 6.5
d. 6.75
e. 6.8

9.

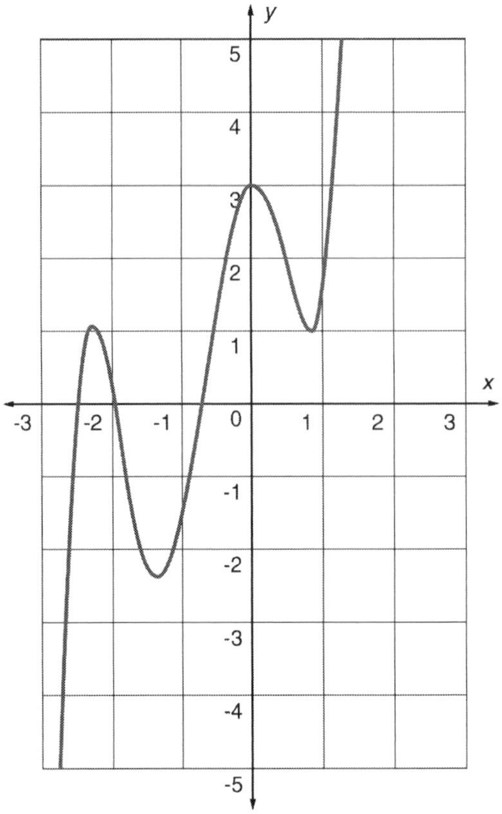


Which of the following functions represents the graph above?

a. $y = x^5 + 3.5x^4 - 6.5x^2 + 0.5x + 3$
b. $y = x^5 - 3.5x^4 + 6.5x^2 - 0.5x - 3$
c. $y = 5x^4 - 6.5x^2 + 0.5x + 3$
d. $y = -5x^4 - 6.5x^2 + 0.5x + 3$
e. $y = 5x^4 + 6.5x^2 - 0.5x - 3$

10. What type of function is modeled by the values in the following table?

| X | f(x) |
|---|---|
| 1 | 2 |
| 2 | 4 |
| 3 | 8 |
| 4 | 16 |
| 5 | 32 |

a. Linear
b. Exponential
c. Quadratic
d. Cubic
e. Logarithmic

11. An investment of \$2,000 is made into an account with an annual interest rate of 5%, compounded continuously. What is the total value for the investment after eight years?
   a. \$2,954.91
   b. \$3,000
   c. \$2,983.65
   d. \$2,800
   e. \$1,977.61

12. What is the y-intercept for $y = x^2 + 3x - 4$?
   a. $y = 4$ and -4
   b. $y = -4$
   c. $y = 4$
   d. $y = 3$ and -3
   e. $y = -3$

13. Katie works at a clothing company and sold 192 shirts over the weekend. $\frac{1}{3}$ of the shirts that were sold were patterned, and the rest were solid. Which mathematical expression would calculate the number of solid shirts Katie sold over the weekend?
   a. $192 \times \frac{1}{3}$
   b. $192 \div \frac{1}{3}$
   c. $192 \times (1 - \frac{1}{3})$
   d. $192 \div 3$
   e. $192 \times (-\frac{1}{3})$

14. What is the function that forms an equivalent graph to $y = \cos(x)$?
   a. $y = \tan(x)$
   b. $y = \csc(x)$
   c. $y = \sin\left(x + \frac{\pi}{2}\right)$
   d. $y = \sin\left(x - \frac{\pi}{2}\right)$
   e. $y = \tan\left(x + \frac{\pi}{2}\right)$

15. Which of the following is the result of simplifying the expression:

$$\frac{4a^{-1}b^3}{a^4b^{-2}} \times \frac{3a}{b}$$

a. $12a^3b^5$

b. $12\frac{b^2}{a^2}$

c. $\frac{12}{a^4}$

d. $7\frac{b^4}{a}$

e. $12\frac{b^4}{a^4}$

16. What are the zeros of the function: $f(x) = x^3 + 4x^2 + 4x$?
a. -2
b. 0, -2
c. 2
d. 0, 2
e. 0

17. What is the solution to the following system of equations?

$$x^2 - 2x + y = 8$$
$$x - y = -2$$

a. $(-2, 3)$
b. There is no solution.
c. $(-2, 0)\ (1, 3)$
d. $(-2, 0)\ (3, 5)$
e. $(2, 0)\ (-1, 3)$

18. How could the following equation be factored to find the zeros?

$$y = x^3 - 3x^2 - 4x$$

a. $0 = x^2(x - 4), x = 0, 4$
b. $0 = 3x(x + 1)(x + 4), x = 0, -1, -4$
c. $0 = x(x + 1)(x + 6), x = 0, -1, -6$
d. $0 = 3x(x + 1)(x - 4), x = 0, 1, -4$
e. $0 = x(x + 1)(x - 4), x = 0, -1, 4$

19. What is the simplified quotient of the following expression?

$$\frac{5x^3}{3x^2y} \div \frac{25}{3y^9}$$

a. $\frac{125x}{9y^{10}}$

b. $\frac{x}{5y^8}$

c. $\frac{5}{xy^8}$

d. $\frac{xy^8}{5}$

e. $\frac{xy^2}{5x}$

20. A line passes through the point (1, 2) and crosses the *y*-axis at *y* = 1. Which of the following is an equation for this line?

a. $y = 2x$
b. $y = x + 1$
c. $x + y = 1$
d. $y = \frac{x}{2} - 2$
e. $y = x - 1$

# Answer Explanations

**1. C:** To solve for the value of b, both sides of the equation need to be equalized.

Start by cancelling out the lower value of -4 by adding 4 to both sides:

$5b - 4 = 2b + 17$
$5b - 4 + 4 = 2b + 17 + 4$
$5b = 2b + 21$

The variable *b* is the same on each side, so subtract the lower 2b from each side:

$5b = 2b + 21$
$5b - 2b = 2b + 21 - 2b$
$3b = 21$

Then divide both sides by 3 to get the value of *b*:

$$3b = 21$$

$$\frac{3b}{3} = \frac{21}{3}$$

$$b = 7$$

**2. E:** This problem involves a composition function, where one function is plugged into the other function. In this case, the $f(x)$ function is plugged into the $g(x)$ function for each $x$-value. The composition equation becomes:

$$g(f(x)) = 2^3 - 3(2^2) - 2(2) + 6$$

Simplifying the equation gives the answer:

$$g(f(x)) = 8 - 3(4) - 2(2) + 6$$

$$8 - 12 - 4 + 6 = -2$$

**3. A:** Simplify this to:

$$(4x^2y^4)^{\frac{3}{2}} = 4^{\frac{3}{2}}(x^2)^{\frac{3}{2}}(y^4)^{\frac{3}{2}}$$

Now:

$$4^{\frac{3}{2}} = (\sqrt{4})^3 = 2^3 = 8$$

For the other, recall that the exponents must be multiplied, so this yields:

$$8x^{2\cdot\frac{3}{2}}y^{4\cdot\frac{3}{2}} = 8x^3y^6$$

**4. B:** The equation can be solved by factoring the numerator into $(x + 6)(x - 5)$. Since that same factor $(x - 5)$ exists on top and bottom, that factor cancels. This leaves the equation $x + 6 = 11$. Solving the equation gives the answer $x = 5$. When this value is plugged into the equation, it yields a zero in the denominator of the fraction. Since this is undefined, there is no solution.

**5. D:** The slope from this equation is 50, and it is interpreted as the cost per gigabyte used. Since the *g*-value represents number of gigabytes and the equation is set equal to the cost in dollars, the slope relates these two values. For every gigabyte used on the phone, the bill goes up 50 dollars.

**6. D:** Using SOHCAHTOA, tangent is $\frac{y}{x}$ for the special triangles. Since the value of $\tan(x)$ needs to be negative one, the angle for the tangent must be some form of 45 degrees or $\frac{\pi}{4}$. The value is negative in the second and fourth quadrant, so the answer is $\frac{3\pi}{4}$ and $\frac{7\pi}{4}$.

**7. A:** The function can be factored to identify the zeros. First, the term $3x$ is factored out to the front because each term contains $3x$. Then, the quadratic is factored into $(x + 3)(x - 2)$.

**8. B:** The independent variable's coordinate at the vertex of a parabola (which is the highest point, when the coefficient of the squared independent variable is negative) is given by:

$$x = -\frac{b}{2a}$$

Substitute and solve for *x* to get:

$$x = -\frac{4}{2(-16)} = \frac{1}{8}$$

Using this value of *x*, the maximum height of the ball (*y*), can be calculated. Substituting *x* into the equation yields:

$$h(t) = -16(\frac{1}{8})^2 + 4(\frac{1}{8}) + 6 = 6.25$$

**9. A:** The graph contains four turning points (where the curve changes from rising to falling or vice versa). This indicates that the degree of the function (highest exponent for the variable) is 5, eliminating Choices *C, D,* and *E*. The $y$-intercepts of the functions can be determined by substituting 0 for $x$ and finding the value of $y$. The function for Choice *A* has a $y$-intercept of 3, and the function for Choice *B* has a $y$-intercept of -3. Therefore, Choice *B* is eliminated.

**10. B:** The table shows values that are increasing exponentially. The differences between the inputs are the same, while the differences in the outputs are changing by a factor of 2. The values in the table can be modeled by the equation $f(x) = 2^x$.

**11. C:** The formula for continually compounded interest is $A = Pe^{rt}$. Plugging in the given values to find the total amount in the account yields the equation:

$$A = 2000e^{0.05*8} = 2983.65$$

**12. B:** The y-intercept of an equation is found where the $x$-value is zero. Plugging zero into the equation for $x$ allows the first two terms to cancel out, leaving -4.

**13. C:** $\frac{1}{3}$ of the shirts sold were patterned. Therefore, $1 - \frac{1}{3} = \frac{2}{3}$ of the shirts sold were solid. Anytime "of" a quantity appears in a word problem, multiplication needs to be used. Therefore:

$$192 \times \frac{2}{3} = 192 \times \frac{2}{3} = \frac{384}{3} = 128 \text{ solid shirts were sold}$$

The entire expression is $192 \times \left(1 - \frac{1}{3}\right)$.

**14. C:** Graphing the function $y = \cos(x)$ shows that the curve starts at $(0, 1)$, has an amplitude of 2, and a period of $2\pi$. This same curve can be constructed using the sine graph, by shifting the graph to the left $\frac{\pi}{2}$ units. This equation is in the form:

$$y = \sin(x + \frac{\pi}{2})$$

**15. E:** To simplify the given equation, the first step is to make all exponents positive by moving them to the opposite place in the fraction. This expression becomes:

$$\frac{4b^3b^2}{a^1a^4} \times \frac{3a}{b}$$

Then the rules for exponents can be used to simplify. Multiplying the same bases means the exponents can be added. Dividing the same bases means the exponents are subtracted.

**16. B:** There are two zeros for the given function. They are $x = 0, -2$. The zeros can be found a number of ways, but this particular equation can be factored into:

$$f(x) = x(x^2 + 4x + 4)$$

$$x(x + 2)(x + 2)$$

By setting each factor equal to zero and solving for $x$, there are two solutions. On a graph, these zeros can be seen where the line crosses the $x$-axis.

**17. D:** This system of equations involves one quadratic function and one linear function, as seen from the degree of each equation. One way to solve this is through substitution. Solving for y in the second equation yields $y = x + 2$. Plugging this equation in for the y of the quadratic equation yields:

$$x^2 - 2x + x + 2 = 8$$

Simplifying the equation, it becomes $x^2 - x + 2 = 8$. Setting this equal to zero and factoring, it becomes:

$$x^2 - x - 6 = 0 = (x - 3)(x + 2)$$

Solving these two factors for *x* gives the zeros $x = 3, -2$. To find the y-value for the point, each number can be plugged in to either original equation. Solving each one for y yields the points $(3, 5)$ and $(-2, 0)$.

**18. E:** Finding the zeros for a function by factoring is done by setting the equation equal to zero, then completely factoring. Since there was a common $x$ for each term in the provided equation, that would be factored out first. Then the quadratic that was left could be factored into two binomials, which are $(x + 1)(x - 4)$. Setting each factor equal to zero and solving for $x$ yields three zeros.

**19. D:** Dividing rational expressions follows the same rule as dividing fractions. The division is changed to multiplication by the reciprocal of the second fraction. This turns the expression into:

$$\frac{5x^3}{3x^2y} \times \frac{3y^9}{25}$$

Multiplying across and simplifying, the final expression is $\frac{xy^8}{5}$.

**20. B:** From the slope-intercept form, $y = mx + b$, it is known that *b* is the *y*-intercept, which is 1. Compute the slope as $\frac{2-1}{1-0} = 1$, so the equation should be $y = x + 1$.

# Geometry and Measurement

## *Coordinate*

### Solving Problems in the Coordinate Plane

The location of a point on a coordinate grid is identified by writing it as an ordered pair. An **ordered pair** is a set of numbers indicating the *x*-and *y*-coordinates of the point. Ordered pairs are written in the form (*x*, *y*) where *x* and *y* are values which indicate their respective coordinates. For example, the point (3, -2) has an *x*-coordinate of 3 and a *y*-coordinate of -2.

Plotting a point on the coordinate plane with a given coordinate means starting from the **origin** (0, 0). To determine the value of the *x*-coordinate, move right (positive number) or left (negative number) along the *x*-axis. Next, move up (positive number) or down (negative number) to the value of the *y*-coordinate. Finally, plot and label the point. For example, plotting the point (1, -2) requires starting from the origin and moving right along the *x*-axis to positive one, then moving down until straight across from negative 2 on the *y*-axis. The point is plotted and labeled. This point, along with three other points, are plotted and labeled on the graph below.

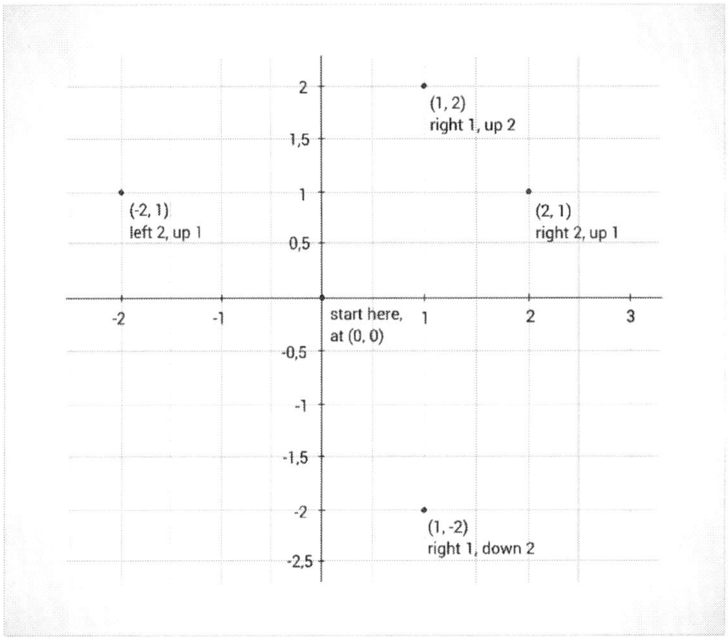


To write the coordinates of a point on the coordinate grid, a line should be traced directly above or below the point until reaching the *x*-axis (noting the value on the *x*-axis). Then, returning to the point, a line should be traced directly to the right or left of the point until reaching the *y*-axis (noting the value on the *y*-axis). The ordered pair (*x*, *y*) should be written with the values determined for the *x*- and *y*-coordinates.

Polygons can be drawn in the coordinate plane given the coordinates of their vertices. These coordinates can be used to determine the perimeter and area of the figure. Suppose triangle *RQP* has vertices located at the points: $R(-4, 2)$, $Q(1, 6)$, and $P(1, 2)$. By plotting the points for the three vertices, the triangle can be constructed as follows:

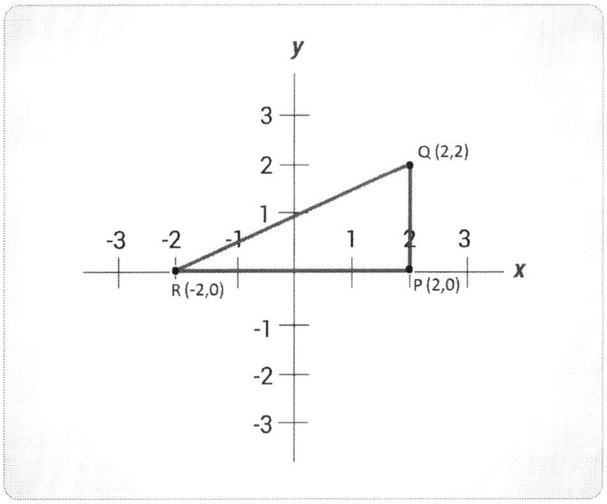


Because points $R$ and $P$ have the same $y$-coordinates (they are directly across from each other), the distance between them is determined by subtracting their $x$-coordinates (or simply counting units from one point to the other): 2 – (-2) = 4. Therefore, the length of side $RP$ is 4 units. Because points $Q$ and $P$ have the same $x$-coordinate (they are directly above and below each other), the distance between them is determined by subtracting their $y$-coordinates (or counting units between them): 2 – 0 = 2. Therefore, the length of side $PQ$ is 2 units. Knowing the length of side $RP$, which is the base of the triangle, and the length of side $PQ$, which is the height of the triangle, the area of the figure can be determined by using the formula $A = \frac{1}{2}bh$.

To determine the perimeter of the triangle, the lengths of all three sides are needed. Points $R$ and $Q$ are neither directly across nor directly above and below each other. Therefore, the distance formula must be used to find the length of side $RQ$. The distance formula is as follows:

$$d = \sqrt{(x_2 - x_1)^2 + (y_2 - y_1)^2}$$

$$d = \sqrt{(2 - (-2))^2 + (2 - 0)^2}$$

$$d = \sqrt{(4)^2 + (2)^2}$$

$$d = \sqrt{16 + 4}$$

$$d = \sqrt{20}$$

The perimeter is determined by adding the lengths of the three sides of the triangle.

## Lines

A line is as series of points that extends in both directions without ending. It consists of an infinite number of points and is drawn with arrows on both ends to indicate it extends infinitely. Lines can be named by two points on the line or with a single, cursive, lower case letter. The lines below are named: line *AB* or line *BA* or $\overleftrightarrow{AB}$ or $\overleftrightarrow{BA}$; and line *m*.

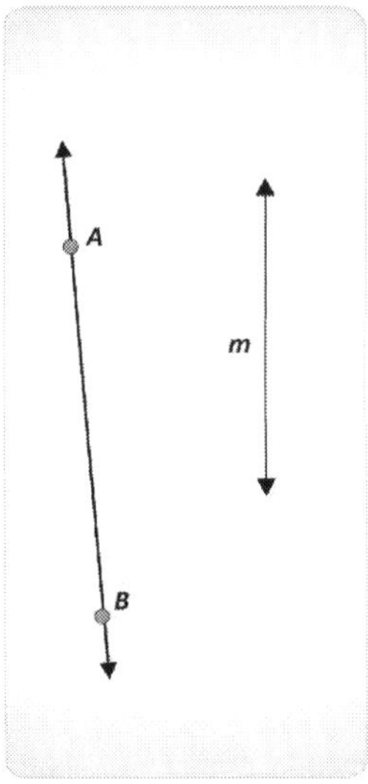


Two lines are considered parallel to each other if, while extending infinitely, they will never intersect (or meet). Parallel lines point in the same direction and are always the same distance apart. Two lines are considered perpendicular if they intersect to form right angles. Right angles are 90°. Typically, a small box is drawn at the intersection point to indicate the right angle.

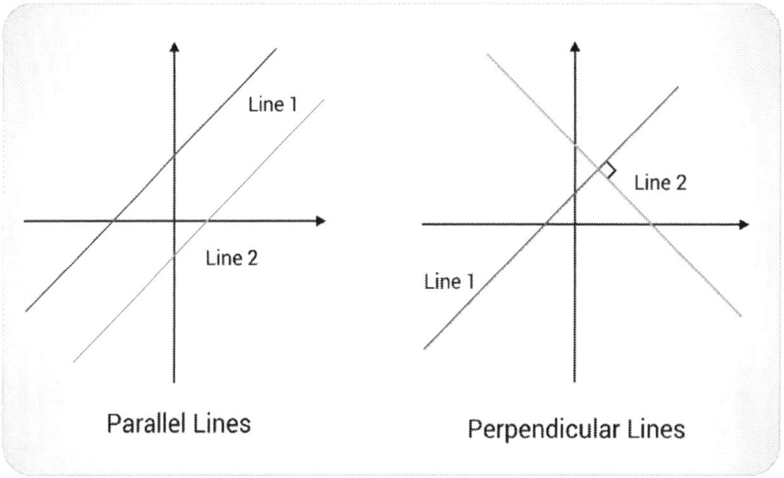

Line 1 is parallel to line 2 in the left image and is written as line 1 | | line 2. Line 1 is perpendicular to line 2 in the right image and is written as line 1 ⊥ line 2.

A ray has a specific starting point and extends in one direction without ending. The endpoint of a ray is its starting point. Rays are named using the endpoint first, and any other point on the ray. The following ray can be named ray *AB* and written $\overrightarrow{AB}$.

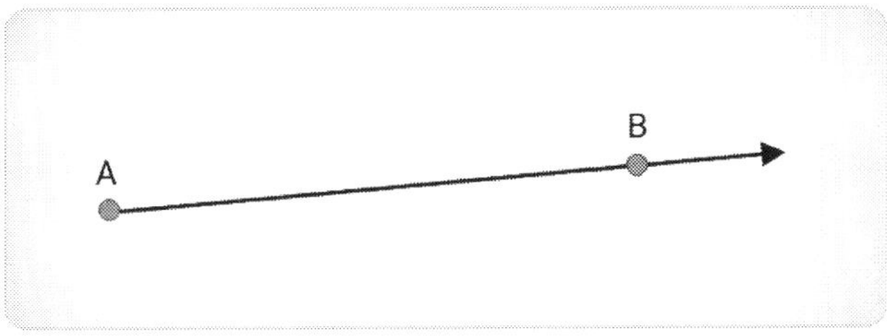


A line segment has specific starting and ending points. A line segment consists of two endpoints and all the points in between. Line segments are named by the two endpoints. The example below is named segment *GH* or segment *HG*, written $\overline{GH}$or $\overline{HG}$.

## Solving Line Problems

Two lines are parallel if they have the same slope and a different intercept. Two lines are perpendicular if the product of their slope equals -1. Parallel lines never intersect unless they are the same line, and perpendicular lines intersect at a right angle. If two lines aren't parallel, they must intersect at one point. Determining equations of lines based on properties of parallel and perpendicular lines appears in word problems. To find an equation of a line, both the slope and a point the line goes through are necessary. Therefore, if an equation of a line is needed that's parallel to a given line and runs through a specified point, the slope of the given line and the point are plugged into the point-slope form of an equation of a line. Secondly, if an equation of a line is needed that's perpendicular to a given line running through a specified point, the negative reciprocal of the slope of the given line and the point are plugged into the point-slope form. Also, if the point of intersection of two lines is known, that point will be used to solve the set of equations. Therefore, to solve a system of equations, the point of intersection must be found. If a set of two equations with two unknown variables has no solution, the lines are parallel.

## Solving Problems with Parallel and Perpendicular Lines

Two lines can be parallel, perpendicular, or neither. If two lines are parallel, they have the same slope. This is proven using the idea of similar triangles. Consider the following diagram with two parallel lines, L1 and L2:

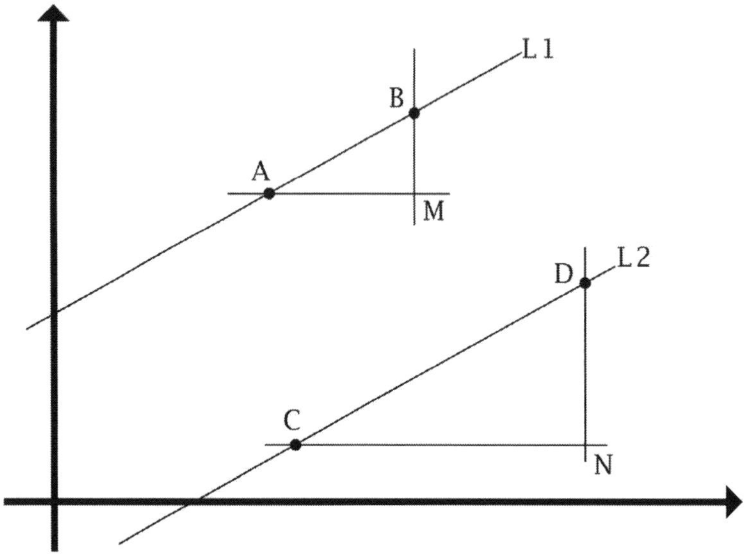


A and B are points on L1, and C and D are points on L2. Right triangles are formed with vertex M and N where lines BM and DN are parallel to the *y*-axis and AM and CN are parallel to the $x$-axis. Because all three sets of lines are parallel, the triangles are similar. Therefore, $\frac{\text{BM}}{\text{DN}} = \frac{\text{MA}}{\text{NC}}$.This shows that the rise/run is equal for lines L1 and L2. Hence, their slopes are equal.

Secondly, if two lines are perpendicular, the product of their slopes equals -1. This means that their slopes are negative reciprocals of each other. Consider two perpendicular lines, *l* and *n*:

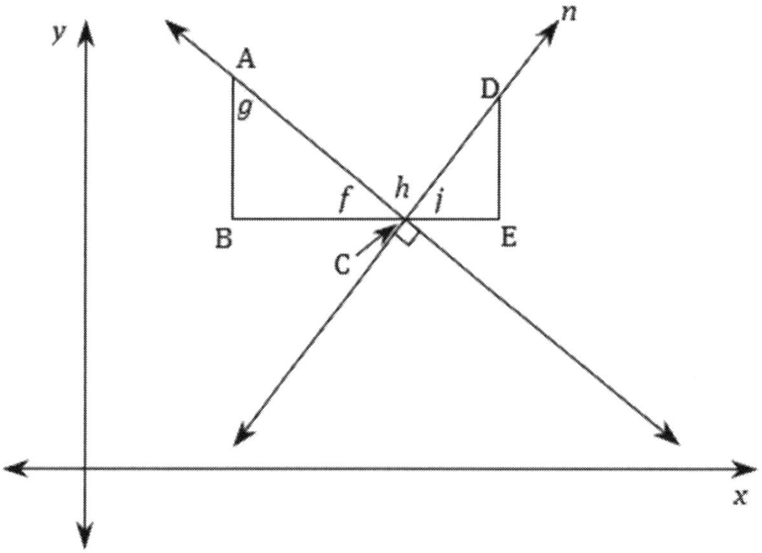

Right triangles ABC and CDE are formed so that lines BC and CE are parallel to the $x$-axis, and AB and DE are parallel to the $y$-axis. Because line BE is a straight line, angles:

$$f + h + i = 180 \; degrees$$

However, angle $h$ is a right angle, so:

$$f + j = 90 \; degrees$$

By construction, $f + g = 90$, which means that $g = j$. Therefore, because angles $B = E$ and $g = j$, the triangles are similar and:

$$\frac{\text{AB}}{\text{BC}} = \frac{\text{CE}}{\text{DE}}$$

Because slope is equal to rise/run, the slope of line $l$ is $-\frac{AB}{BC}$ and the slope of line $n$ is $\frac{DE}{CE}$. Multiplying the slopes together gives:

$$-\frac{AB}{BC} \cdot \frac{DE}{CE} = -\frac{CE}{DE} \cdot \frac{DE}{CE} = -1$$

This proves that the product of the slopes of two perpendicular lines equals -1. Both parallel and perpendicular lines can be integral in many geometric proofs, so knowing and understanding their properties is crucial for problem-solving.

## Proving Theorems About Lines and Angles

To prove any geometric theorem, the proven theorems must be linked in a logical order that flows from an original point to the desired result. Proving theorems about lines and angles is the basis of proving theorems that involve other shapes. A **transversal** is a line that passes through two lines at two points. Common theorems that need to be proved are: vertical angles are congruent; a transversal passing through two parallel lines forms two alternate interior angles that are congruent and two corresponding angles that are congruent; and points on a perpendicular bisector of a line segment are equidistant from the endpoints of the line segment.

## Proving Theorems with Coordinates

Many important formulas and equations exist in geometry that use coordinates. The distance between two points $(x_1, y_1)$ and $(x_2, y_2)$ is:

$$d = \sqrt{(x_2 - x_1)^2 + (y_2 - y_1)^2}$$

The slope of the line containing the same two points is:

$$m = \frac{y_2 - y_1}{x_2 - x_1}$$

Also, the midpoint of the line segment with endpoints $(x_1, y_1)$ and $(x_2, y_2)$ is:

$$M = \left(\frac{x_1 + x_2}{2}, \frac{y_1 + y_2}{2}\right)$$

The equations of a circle, parabola, ellipse, and hyperbola can also be used to prove theorems algebraically. Knowing when to use which formula or equation is extremely important, and knowing which formula applies to which property of a given geometric shape is an integral part of the process. In some cases, there are a number of ways to prove a theorem; however, only one way is required.

## Formulas for Ratios

If a line segment with endpoints $(x_1, y_1)$ and $(x_2, y_2)$ is partitioned into two equal parts, the formula for midpoint is used. Recall this formula is $M = \left(\frac{x_1+x_2}{2}, \frac{y_1+y_2}{2}\right)$ and the ratio of line segments is 1:1. However, if the ratio needs to be anything other than 1:1, a different formula must be used. Consider a ratio that is $a: b$. This means the desired point that partitions the line segment is $\frac{a}{a+b}$ of the way from $(x_1, y_1)$ to $(x_2, y_2)$. The actual formula for the coordinate is:

$$\left(\frac{bx_1 + ax_2}{a+b} \frac{by_1 + ay_2}{a+b}\right)$$

## Computing Side Length, Perimeter, and Area

The side lengths of each shape can be found by plugging the endpoints into the distance formula $d = \sqrt{(x_2 - x_1)^2 + (y_2 - y_1)^2}$ between two ordered pairs $(x_1, y_1)$ and $(x_2, y_2)$. The distance formula is derived from the Pythagorean theorem. Once the side lengths are found, they can be added together to obtain the perimeter of the given polygon. Simplifications can be made for specific shapes such as squares and equilateral triangles. For example, one side length can be multiplied by 4 to obtain the perimeter of a square.

Also, one side length can be multiplied by 3 to obtain the perimeter of an equilateral triangle. A similar technique can be used to calculate areas. For polygons, both side length and height can be found by using the same distance formula. Areas of triangles and quadrilaterals are straightforward through the use of $A = \frac{1}{2}bh$ or $A = bh$, depending on the shape. To find the area of other polygons, their shapes can be partitioned into rectangles and triangles. The areas of these simpler shapes can be calculated and then added together to find the total area of the polygon.

### Relationships between Angles

**Supplementary angles** add up to 180 degrees. **Vertical angles** are two nonadjacent angles formed by two intersecting lines. **Corresponding angles** are two angles in the same position whenever a straight line (known as a **transversal**) crosses two others. If the two lines are parallel, the corresponding angles are equal. **Alternate interior angles** are also a pair of angles formed when two lines are crossed by a transversal. They are opposite angles that exist inside of the two lines. In the corresponding angles diagram below, angles 2 and 7 are alternate interior angles, as well as angles 6 and 3. **Alternate exterior angles** are opposite angles formed by a transversal but, in contrast to interior angles, exterior angles exist outside the two original lines. Therefore, angles 1 and 8 are alternate exterior angles and so are angles 5 and 4. Finally, **consecutive interior angles** are pairs of angles formed by a transversal. These angles are located on the same side of the transversal and inside the two original lines. Therefore, angles 2 and 3 are a pair of consecutive interior angles, and so are angles 6 and 7. These definitions are instrumental in solving many problems that involve determining relationships between angles.

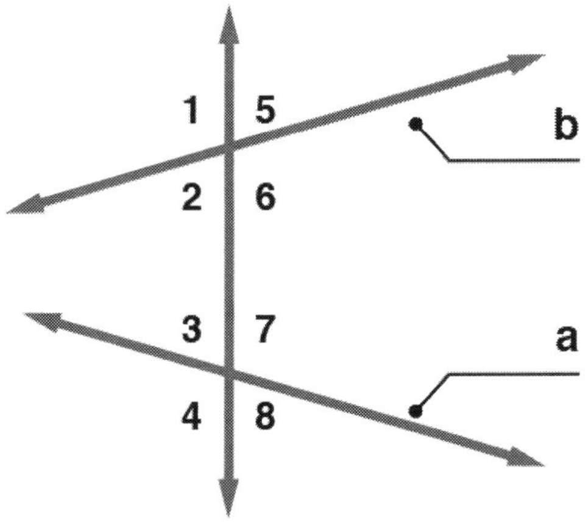


Corresponding Angles

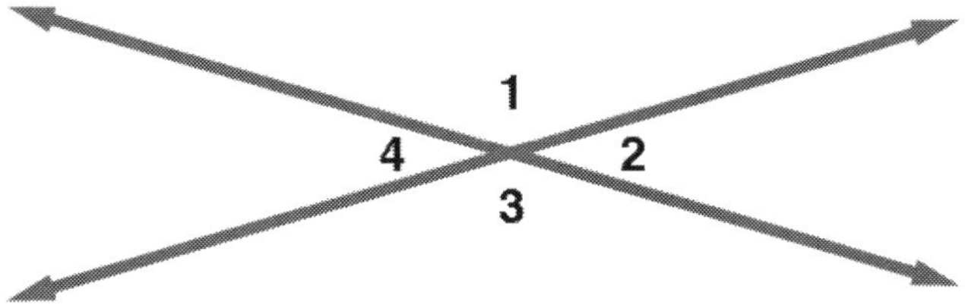


Vertical Angles

## Sets of Points in the Plane

The **midpoint** between two points, $(x_1, y_1)$ and $(x_2, y_2)$, is given by taking the average of the x coordinates and the average of the y coordinates:

$$\left(\frac{x_1 + x_2}{2}, \frac{y_1 + y_2}{2}\right)$$

The **distance** between two points, $(x_1, y_1)$ and $(x_2, y_2)$, is given by the **Pythagorean formula:**

$$\sqrt{(x_2 - x_1)^2 + (y_2 - y_1)^2}$$

To find the perpendicular distance between a line $Ax + By = C$ and a point $(x_1, y_1)$ not on the line, we need to use this formula:

$$\frac{|Ax_1 + By_1 + C|}{\sqrt{A^2 + B^2}}$$

# Parabolas

The graph of an equation of the form $y = ax^2 + bx + c$ or $x = ay^2 + by + c$ is called a **parabola.** A parabola is defined as a specific type of curve such that any point on it is the same distance from a fixed point (called the **foci**) and a fixed straight line (called the **directrix**). A parabola is the shape formed from the intersection of a cone with a plane that's parallel to its side. Every parabola has an **axis of symmetry,** and its vertex $(h, k)$ is the point at which the axis of symmetry intersects the curve. If the parabola has an axis of symmetry parallel to the y-axis, the focus is the point $(h, k + f)$ and the directrix is the line $y = k - f$. For example, a parabola may have a vertex at the origin, focus $(0, f)$, and directrix $y = -f$. The equation of this parabola can be derived by using both the focus and the directrix. The distance from any coordinate on the curve to the focus is the same as the distance to the directrix, and the Pythagorean theorem can be used to find the length of *d.* The triangle has sides with length $|x|$ and $|y - f|$ and therefore:

$$d = \sqrt{x^2 + (y - f)^2}$$

By definition, the vertex is halfway between the focus and the directrix and $d = y + f$. Setting these two equations equal to one another, squaring each side, simplifying, and solving for *y* gives the equation of a parabola with the focus *f* and the vertex being the origin:

$$y = \frac{1}{4f}x^2$$

If the vertex $(h, k)$ is not the origin, a similar process can be completed to derive the equation $(x-h)^2 = 4f(y-k)$ for a parabola with focus $f$.

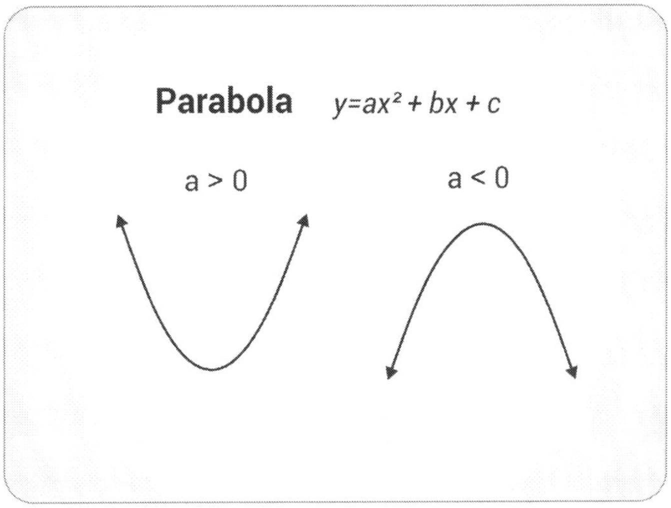


## Circles

### Equation of a Circle

A **circle** can be defined as the set of all points that are the same distance (known as the **radius, r**) from a single point *C* (known as the **center** of the circle). The center has coordinates $(h, k)$, and any point on the circle can be labelled with coordinates $(x, y)$. As shown below, a **right triangle** is formed with these two points:

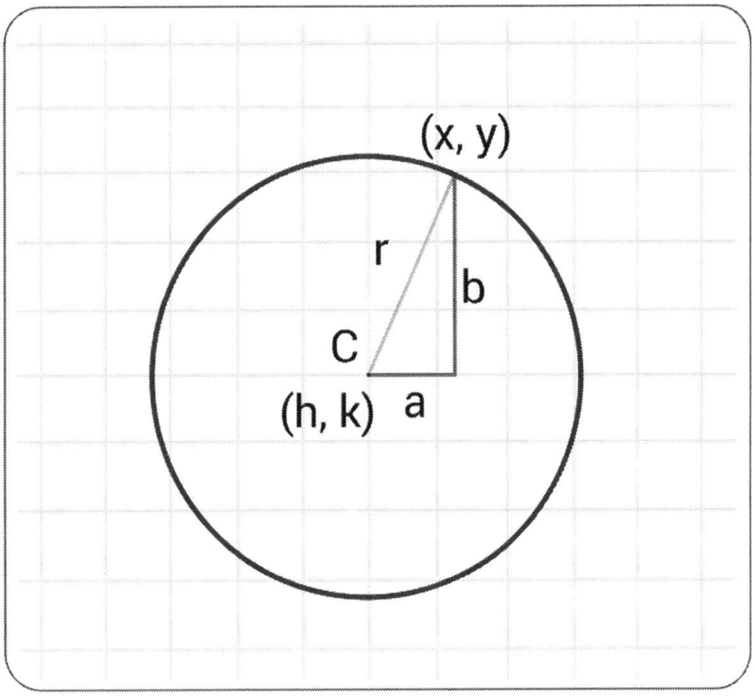

The Pythagorean theorem states that:

$$a^2 + b^2 = r^2$$

However, *a* can be replaced by $|x - h|$ and *b* can be replaced by $|y - k|$ by using the **distance formula**. That substitution results in $(x - h)^2 + (y - k)^2 = r^2$, which is the formula for finding the equation of any circle with a center $(h, k)$ and a radius *r*. Note that sometimes *c* is used instead of *r*.

## Circles in the Coordinate Plane

If a circle is placed on the coordinate plane with the center of the circle at the origin (0,0), then point (*x*, *y*) is a point on the circle. Furthermore, the line extending from the center to point (*x*, *y*) is the radius, or *r*. By applying the Pythagorean Theorem ($a^2 + b^2 = c^2$), it can be stated that $x^2 + y^2 = r^2$. However, the center of the circle does not always need to be on the origin of the coordinate plane.

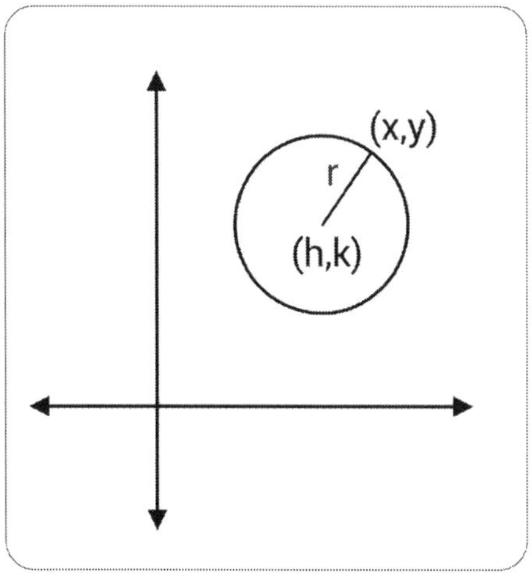


In the diagram above, the center of the circle is noted by (*h*, *k*). By applying the distance formula, the equation becomes:

$\sqrt{(x - h)^2 + (y - k)^2}$

When squaring both sides of the equation, the result is the standard form of a circle with the center (*h*, *k*) and radius *r*. Namely, $r^2 = (x - h)^2 + (y - k)^2$, where *r* = radius and center = (*h*, *k*). The following examples may be solved by using this information:

Example: Graph the equation $-x^2 + y^2 = 25$

To graph this equation, first note that the center of the circle is (0, 0). The radius is the positive square root of 25 or 5.

Example: Find the equation for the circle below.

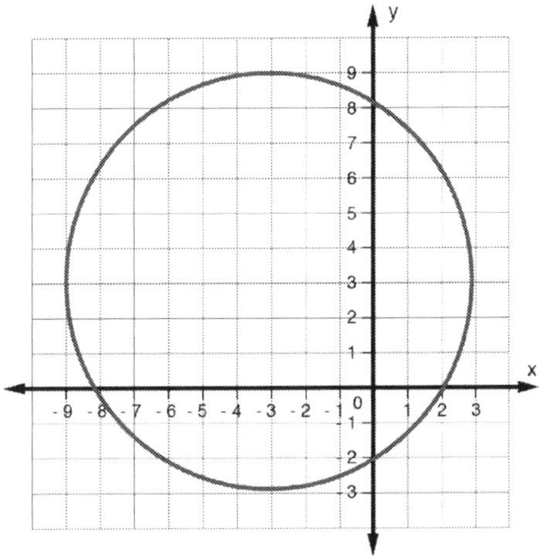


To find the equation for the circle, note that its center is not zero. Therefore, to find the circle's center, draw vertical and horizontal diameters to examine where they intersect. The center is located at point: (-3, 3). Next, count the number of spaces from the center to the outside of the circle. This number is 6. Therefore, 6 is the radius. Finally, plug in the numbers that are known into the standard equation for a circle:

$$36 = \left(x - (-3)\right)^2 + (y - 3)^2$$

$$or$$

$$36 = (x + 3)^2 + (y - 3)^2$$

It is possible to determine whether a point lies on a circle or not within the coordinate plane. For example, a circle has a center of (2, -5), and a radius of 6 centimeters. The first step is to apply the equation of a circle, which is $r^2 = (x - h)^2 + (y - k)^2$, where *r* = radius and the center = (*h*, *k*). Next, substitute the numbers for the center point and the number for the radius. This action simplifies the equation to:

$$36 = (x - 2)^2 + (y + 5)^2$$

Note that the radius of 6 was squared to get 36.

To prove that the point (2, -1) lies on the circle, apply the equation of the circle that was just used and input the values of (2, -1) for *x* and *y* in the equation.

$$36 = (x - 2)^2 + (y + 5)^2$$

$$36 = (2 - 2)^2 + (1 + 5)^2$$

$$36 = (0)^2 + (6)^2$$

$$36 = 36$$

Because the left side of the equation equals the right side of the equation, point (2, 1) lies on the given circle.

## Solving Basic Problems Involving Central Angles, Tangents, Arcs, and Sectors

To find any fractional area of a circle, the formula $A = \frac{C}{360}\pi r^2$ can be used, where *C* is the number of degrees of the central angle of the sector. The area of a circle can also be found by using the arc length rather than the degree of the sector. This formula is $A = rs^2$, where *s* is the arc length and *r* is the radius of the circle.

A **chord** of a circle is a straight line formed when its endpoints are allowed to be any two points on the circle. If the radius and the value of the angle subtended at the center by the chord is known, the formula to find the chord length is:

$$C = 2 \times \textit{radius} \times \sin\frac{\textit{angle}}{2}$$

Remember that this formula is based on half the length of the chord, so the radius is doubled to determine the full length of the chord.

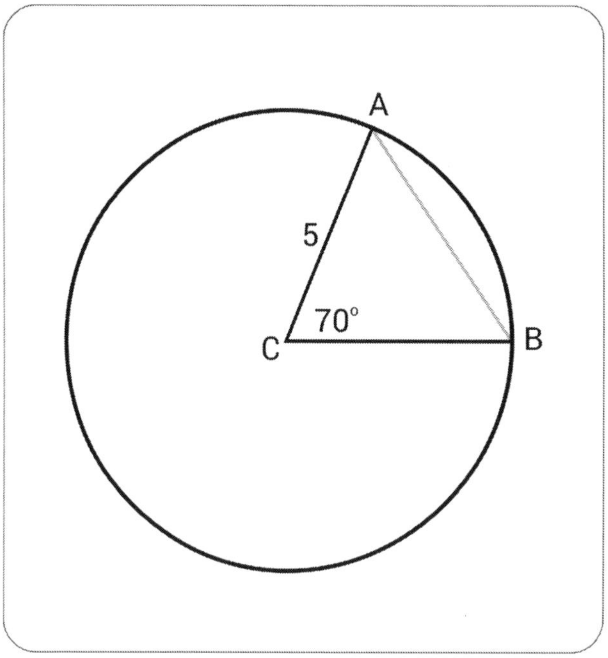


For example, the radius in the diagram above is 5 and the angle is 70 degrees. To find the chord length, plug in the values for the radius and angle to obtain the answer of 5.7.

$$5 \times \sin\frac{70}{2} = 2.87 \times 2 = 5.7$$

Many angles exist within a circle. A **central angle** is formed by using two radii as its rays and the center of the circle as its vertex. An inscribed angle is formed by using two chords as its rays, and its vertex is a point on the circle itself. Finally, a **circumscribed angle** has a vertex that is a point outside the circle and rays that intersect with the circle. Some relationships exist between these types of angles, and, in order to define these relationships, arc measure must be understood. An **arc** of a circle is a portion of the circumference. Finding the **arc measure** is the same as finding the degree measure of the central angle that intersects the circle to form the arc. The measure of an inscribed angle is half the measure of its intercepted arc. It's also true that the measure of a circumscribed angle is equal to 180 degrees minus the measure of the central angle that forms the arc in the angle.

A **tangent line** is a line that touches a curve at a single point without going through it. A **compass** and a **straight edge** are the tools necessary to construct a tangent line from a point *P* outside the circle to the circle. A tangent line is constructed by drawing a line segment from the center of the circle *O* to the point *P*, and then finding its **midpoint** *M* by bisecting the line segment. By using *M* as the center, a compass is used to draw a circle through points *O* and *P*. *N* is defined as the intersection of the two circles. Finally, a line segment is drawn through *P* and *N*. This is the tangent line. Each point on a circle has only one tangent line, which is perpendicular to the radius at that point. A line similar to a tangent line is a **secant line**. Instead of intersecting the circle at one point, a secant line intersects the circle at two points. A **chord** is a smaller portion of a secant line.

Here's an example:

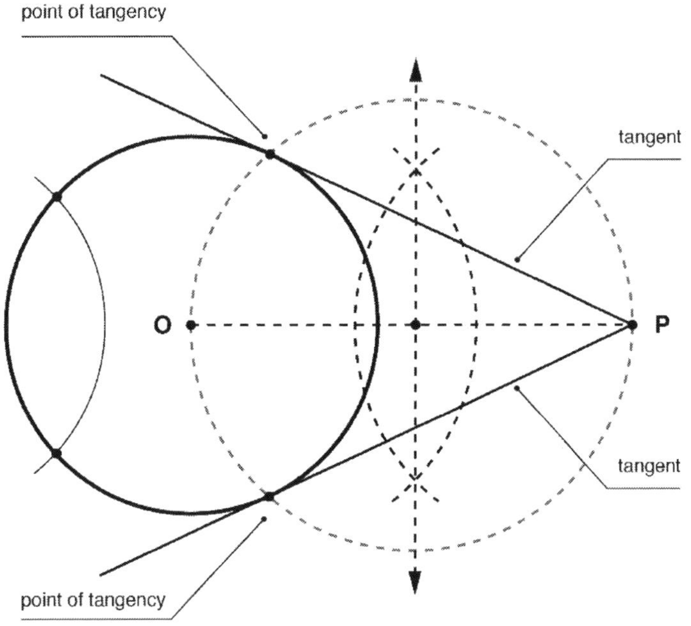


## Arc Length

As previously mentioned, angles can be measured in radians, and 180 degrees equals π radians. Therefore, the measure of a complete circle is $2\pi$ radians. In addition to arc measure, **arc length** can also be found because the length of an arc is a portion of the circle's circumference. The following proportion is true

$$\frac{\text{Arc measure}}{\text{360 degrees}} = \frac{\text{arc length}}{\text{arc circumference}}$$

Arc measure is the same as the measure of the central angle, and this proportion can be rewritten as:

$$\text{arc length} = \frac{\text{central angle}}{\text{360 degrees}} \times \text{circumference}$$

In addition, the degree measure can be replaced with radians to allow the use of both units. Note that arc length is a fractional part of circumference because:

$$\frac{\text{central angle}}{360\text{ degrees}} < 1$$

## Area of a Sector

A **sector** of a circle is a portion of the circle that's enclosed by two radii and an arc. It resembles a piece of a pie, and the area of a sector can be derived using known definitions. The area of a circle can be calculated using the formula $A = \pi r^2$, where *r* is the radius of the circle. The area of a sector of a circle is a fraction of that calculation. For example, if the central angle $\theta$ is known in radians, the area of a sector is defined as:

$$A_s = \pi r^2 \frac{\theta}{2\pi} = \frac{\vartheta r^2}{2}$$

If the angle $\theta$ in degrees is known, the area of the sector is:

$A_s = \frac{\vartheta \pi r^2}{360}$ Finally, if the arc length *L* is known, the area of the sector can be reduced to $A_s = \frac{rL}{2}$.

## Quadrilateral Angles

If a quadrilateral is inscribed in a circle, the sum of its opposite angles is 180 degrees. Consider the quadrilateral ABCD centered at the point O:

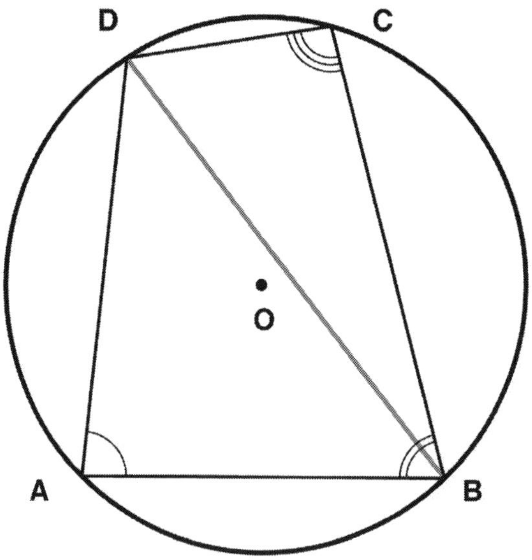


Each of the four line segments within the quadrilateral is a chord of the circle. Consider the diagonal DB. Angle DAB is an inscribed angle leaning on the arc DCB. Therefore, angle DAB is half the measure of the arc DCB. Conversely, angle DCB is an inscribed angle leaning on the arc DAB. Therefore, angle DCB is half the measure of the arc DAB. The sum of arcs DCB and DAB is 360 degrees because they make up the entire circle. Therefore, the sum of angles DAB and DCB equals half of 360 degrees, which is 180 degrees.

## Ellipses

The graph of an equation of the form $\frac{(x-x_0)^2}{a^2}+\frac{(y-y_0)^2}{b^2}=1$ is called an **ellipse**. If $a=b$ then this is a circle with **radius** $r=\frac{1}{a}$. An ellipse is the set of all points for which the sum of the distances from two fixed points (known as the *foci*) is constant. The **distance formula** can be used to derive the formulas of both an ellipse and a hyperbola, given the coordinates of the foci. Consider an ellipse where its major axis is horizontal (i.e., it's longer along the $x$-axis) and its foci are the coordinates $(-c,0)$ and $(c,0)$. The distance from any point $(x,y)$ to $(-c,0)$ is $d_1=\sqrt{(x+c)^2+y^2}$, and the distance from the same point $(x,y)$ to $(c,0)$ is:

$$d_1=\sqrt{(x-c)^2+y^2}$$

Using the definition of an ellipse, it's true that the sum of the distances from the vertex *a* to each foci is equal to $d_1+d_2$. Therefore, $d_1+d_2=(a+c)+(a-c)=2a$ and:

$$\sqrt{(x+c)^2+y^2}+\sqrt{(x-c)^2+y^2}=2a$$

After a series of algebraic steps, this equation can be simplified to $\frac{x^2}{a^2}+\frac{y^2}{b^2}=1$, which is the equation of an ellipse with a horizontal major axis. In this case, $a>b$. When the ellipse has a vertical major axis, similar techniques result in $\frac{x^2}{b^2}+\frac{y^2}{a^2}=1$, and $a>b$.

## Hyperbolas

The graph of an equation of the form $\frac{x^2}{a^2}-\frac{y^2}{b^2}=1$ or $-\frac{x^2}{a^2}+\frac{y^2}{b^2}=1$ is called a **hyperbola**. A hyperbola is the set of all points for which the difference between the distances from two fixed points (also known as the **foci**) is constant.

The equation of a hyperbola can be derived in a fashion similar to that used to derive the equation of an ellipses. Consider a hyperbola with a horizontal major axis and its foci are also the coordinates $(-c,0)$ and $(c,0)$. Again, the distance from any point $(x,y)$ to $(-c,0)$ is $d_1=\sqrt{(x+c)^2+y^2}$ and the distance from the same point $(x,y)$ to $(c,0)$ is:

$$d_1=\sqrt{(x-c)^2+y^2}$$

Using the definition of a hyperbola, it's true that the difference of the distances from the vertex *a* to each foci is equal to $d_1-d_2$. Therefore:

$$d_1-d_2=(c+a)-(c-a)=2a$$

This means that:

$$\sqrt{(x+c)^2+y^2}-\sqrt{(x-c)^2+y^2}=2a$$

After a series of algebraic steps, this equation can be simplified to $\frac{x^2}{a^2}-\frac{y^2}{b^2}=1$, which is the equation of a hyperbola with a horizontal major axis. In this case, $a>b$. Similar techniques result in the equation $\frac{x^2}{b}-\frac{y^2}{a^2}=1$, where $a>b$, when the hyperbola has a vertical major axis.

## Symmetry

If an object can undergo various transformations (discussed below) and not appear to have changed, then the figure is symmetrical. If an object can be split in half by a line and flipped over that line to lie directly on top of itself, it is said to have **line symmetry**. An example of both types of figures is seen below.

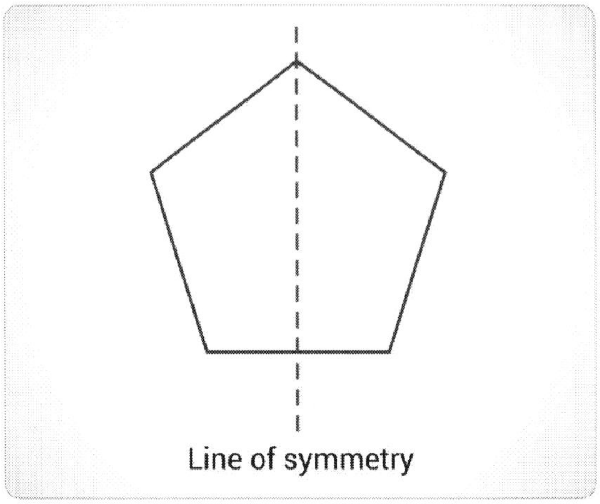


If an object can be rotated about its center to any degree smaller than 360, and it lies directly on top of itself, the object is said to have **rotational symmetry**. An example of this type of symmetry is shown below. The pentagon has an order of 5.

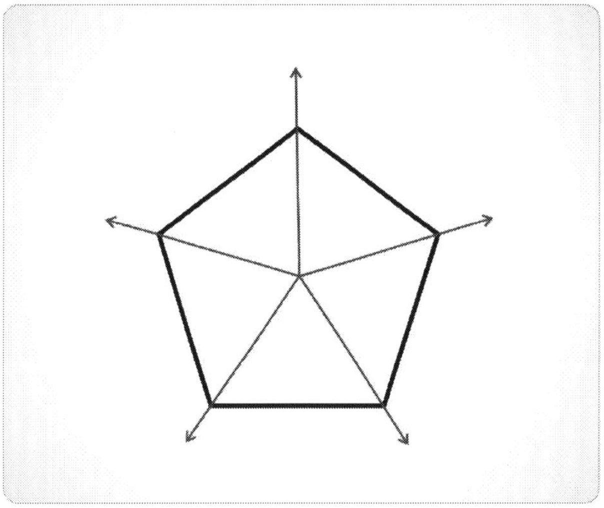

The rotational symmetry lines in the figure above can be used to find the angles formed at the center of the pentagon. Knowing that all of the angles together form a full circle, at 360 degrees, the figure can be split into 5 angles equally. By dividing the 360° by 5, each angle is 72°.

Given the length of one side of the figure, the perimeter of the pentagon can also be found using rotational symmetry. If one side length was 3 cm, that side length can be rotated onto each other side length four times. This would give a total of 5 side lengths equal to 3 cm. To find the perimeter, or distance around the figure, multiply 3 by 5. The perimeter of the figure would be 15 cm.

If a line cannot be drawn anywhere on the object to flip the figure onto itself or rotated less than or equal to 180 degrees to lay on top of itself, the object is asymmetrical. Examples of these types of figures are shown below.

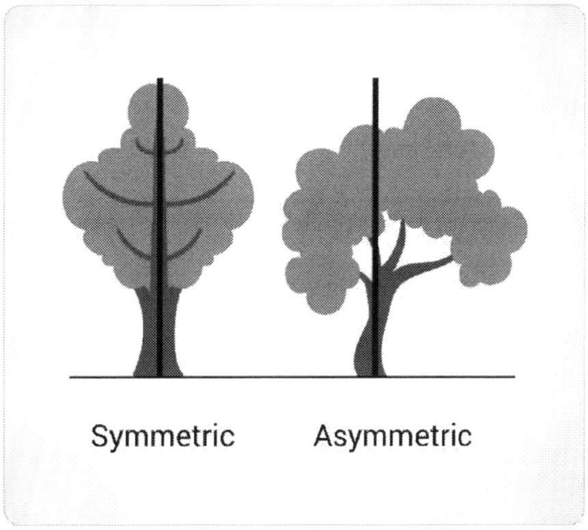

## Transformations of a Plane

Given a figure drawn on a plane, many changes can be made to that figure, including rotation, translation, and reflection. **Rotations** turn the figure about a point, **translations** slide the figure, and **reflections** flip the figure over a specified line. When performing these transformations, the original figure is called the **pre-image**, and the figure after transformation is called the **image.**

More specifically, **translation** means that all points in the figure are moved in the same direction by the same distance. In other words, the figure is slid in some fixed direction. Of course, while the entire figure is slid by the same distance, this does not change any of the measurements of the figures involved. The result will have the same distances and angles as the original figure.

In terms of Cartesian coordinates, a translation means a shift of each of the original points $(x, y)$ by a fixed amount in the *x* and *y* directions, to become $(x + a, y + b)$.

Another procedure that can be performed is called **reflection.** To do this, a line in the plane is specified, called the **line of reflection**. Then, take each point and flip it over the line so that it is the same distance from the line but on the opposite side of it. This does not change any of the distances or angles involved, but it does reverse the order in which everything appears.

To reflect something over the *x*-axis, the points $(x, y)$ are sent to $(x, -y)$. To reflect something over the *y*-axis, the points $(x, y)$ are sent to the points $(-x, y)$. Flipping over other lines is not something easy to express in Cartesian coordinates. However, by drawing the figure and the line of reflection, the distance to the line and the original points can be used to find the reflected figure.

Example: Reflect this triangle with vertices (-1, 0), (2, 1), and (2, 0) over the *y*-axis. The pre-image is shown below.

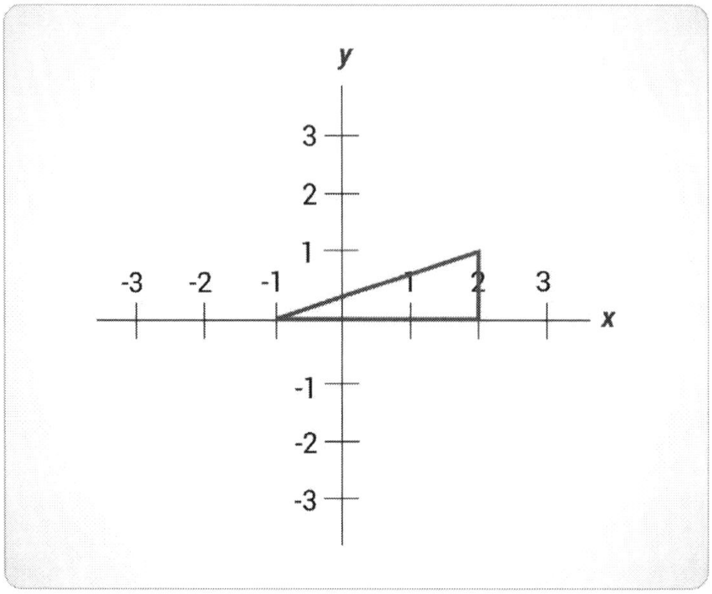


To do this, flip the *x* values of the points involved to the negatives of themselves, while keeping the *y* values the same. The image is shown here.

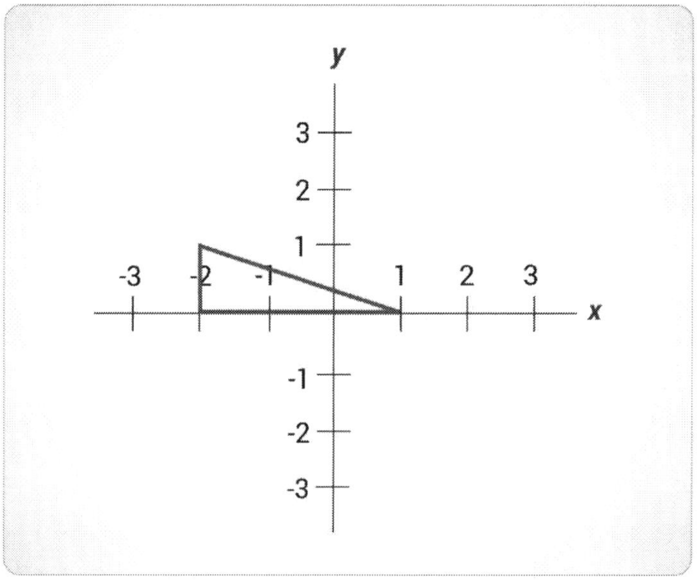


The new vertices will be (1, 0), (-2, 1), and (-2, 0).

Another procedure that does not change the distances and angles in a figure is **rotation.** In this procedure, pick a center point, then rotate every vertex along a circle around that point by the same angle. This procedure is also not easy to express in Cartesian coordinates, and this is not a requirement on this test. However, as with reflections, it's helpful to draw the figures and see what the result of the rotation would look like. This transformation can be performed using a compass and protractor.

Each one of these transformations can be performed on the coordinate plane without changes to the original dimensions or angles.

If two figures in the plane involve the same distances and angles, they are called **congruent figures**. In other words, two figures are congruent when they go from one form to another through reflection, rotation, and translation, or a combination of these.

Remember that rotation and translation will give back a new figure that is identical to the original figure, but reflection will give back a mirror image of it.

To recognize that a figure has undergone a rotation, check to see that the figure has not been changed into a mirror image, but that its orientation has changed (that is, whether the parts of the figure now form different angles with the *x* and *y* axes).

To recognize that a figure has undergone a translation, check to see that the figure has not been changed into a mirror image, and that the orientation remains the same.

To recognize that a figure has undergone a reflection, check to see that the new figure is a mirror image of the old figure.

The following figure is an example of translation, rotation, dilation, and reflection:

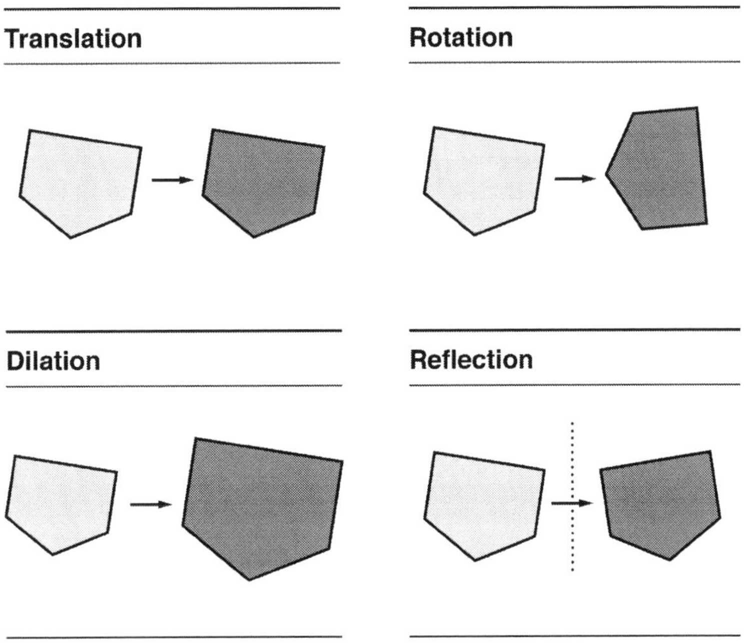


Keep in mind that sometimes a combination of translations, reflections, and rotations may be performed on a figure.

## Distance and Angle Measure

The three major types of transformations (rotations, reflections, and translations) preserve distance and angle measurement. The shapes stay the same, but they are moved to another place in the plane. Therefore, the distance between any two points on the shape doesn't change. Also, any original angle measure between two line segments doesn't change. However, there are transformations that don't preserve distance and angle measurements, including those that don't preserve the original shape. For example, transformations that involve stretching and shrinking shapes don't preserve distance and angle measures. In these cases, the input variables are multiplied by either a number greater than one (**stretch**) or less than one (**shrink**).

## Solving Problems Using Transformations

Most problems in geometry consist of using either congruent or similar shapes and the proofs that follow. The criteria needed to prove triangles are congruent involve both angle and side congruence, which can be proven through properties of transformations. Both pairs of related angles and sides need to be of the same measurement to prove congruence, meaning one triangle can be obtained from the other via reflections, translations, and rotations. The criteria for triangles to be similar involve proportionality of side lengths. Angles are congruent in similar triangles; however, corresponding side lengths only need to be a constant multiple of each other. Relationships in geometric figures other than triangles can be proven using triangle congruence and similarity. If a similar or congruent triangle can be found within another type of geometric figure, their criteria can be utilized to prove a relationship about a given formula. For instance, a rectangle can be broken up into two congruent triangles.

### Rotations and Reflections

A **point of symmetry** is used to determine translations that map a shape onto itself. When a line is drawn through a point of symmetry, it crosses the shape on one side of the point while crossing the shape on the other side at the exact same distance. A shape can be rotated 180 degrees around a point of symmetry to get back to its original shape. Simple examples are the center of a circle and a square. A **line of symmetry** is a line that a shape is folded over to have its sides align, and it goes through the point of symmetry. A combination of reflecting an original shape around a line of symmetry and rotating it can map the original image onto itself. A shape has **rotational symmetry** if it can be rotated to get back to its original shape, and it has **reflection symmetry** if it can be reflected onto itself over some line. Shapes such as rectangles, parallelograms, trapezoids, and regular polygons can be mapped onto themselves through such rotations and reflections.

### Isometric Transformations

Rotations, reflections, and translations are isometric transformations, because throughout each transformation the distance of line segments is maintained, the angle measure is maintained, parallel lines in the original shape remain parallel, and points on lines remain on those lines. A rotation turns a shape around a specific point (*O*) known as the **center of rotation**. An **angle of rotation** is formed by drawing a ray from the center of rotation to a point (*P*) on the original shape and to the point's image (*P'*) on the reflected shape. Thus, it's true that $OP=OP'$. A reflection over a line (*l*), known as the **line of reflection,** takes an original point P and maps it to its image P' on the opposite side of *l*. The line of reflection is the perpendicular bisector of every line formed by an original point and its image. A translation maps each point P in the original shape to a new point P'. The line segment formed between each point and its image consists of the same length, and the line segment formed by two original points is parallel to the line segment formed from their two images.

## Transformation Figures

Once the major transformations are defined, a shape can be altered by performing given transformations. An image is determined from a pre-image by carrying out a series of rotations, reflections, and translations. Once transformations are understood completely, moving and manipulating figures can be related to real-word situations. The transformed figure can then be drawn using either a pencil and paper or geometry software.

## Transformation Mapping

Given a pre-image and an image, transformations can be determined that turn one shape into the other. A sequence of rotations, reflections, and translations can be defined that map the pre-image onto the other shape.

## Dilation

A **dilation** is a transformation that preserves angles, but not distances. This can be thought of as stretching or shrinking a figure. If a dilation makes figures larger, it is called an **enlargement.** If a dilation makes figures smaller, it is called a **reduction.** The easiest example is to dilate around the origin. In this case, multiply the *x* and *y* coordinates by a **scale factor**, $k$, sending points $(x, y)$ to $(kx, ky)$.

As an example, draw a dilation of the following triangle, whose vertices will be the points (-1, 0), (1, 0), and (1, 1).

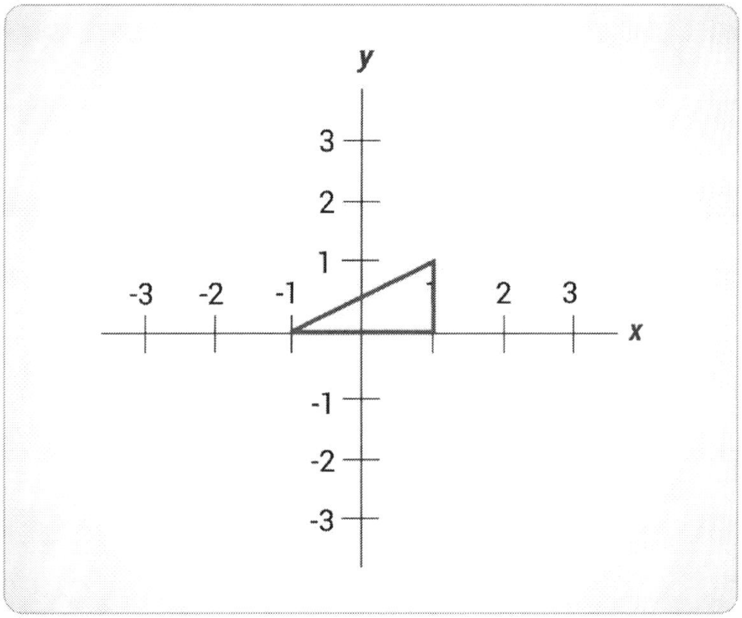

For this problem, dilate by a scale factor of 2, so the new vertices will be (-2, 0), (2, 0), and (2, 2).

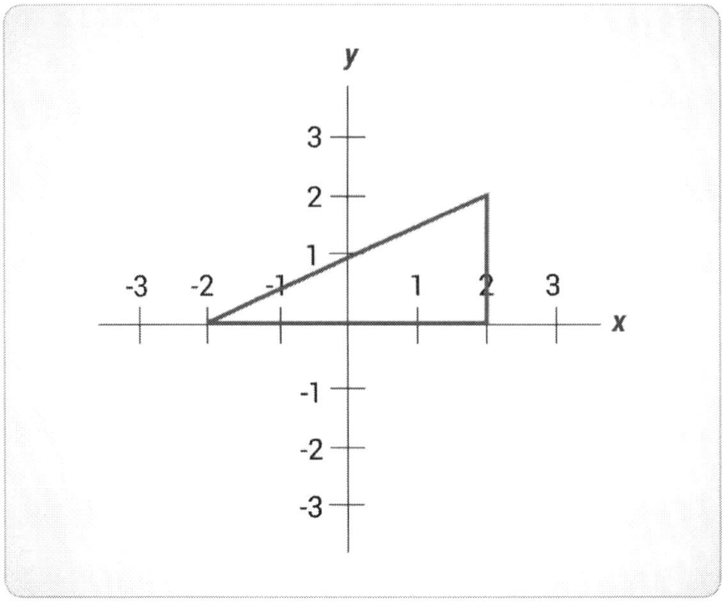


Note that after a dilation, the distances between the vertices of the figure will have changed, but the angles remain the same. The two figures that are obtained by dilation, along with possibly translation, rotation, and reflection, are all **similar** to one another. Another way to think of this is that similar figures have the same number of vertices and edges, and their angles are all the same. Similar figures have the same basic shape, but are different in size.

Consider a rectangle that has been plotted with four corners being the following coordinates: (-4,0), (3,0), (3,-3), and (-4,-3). Now, consider a dilation with a scale factor of 2, meaning all 4 sides will double their length. The resulting is an enlargement of the original image. Their new coordinates of the corners of the rectangle will be (-4,0), (6,6), (6,-6), and (-8,-6). Notice that within all four new corners, each coordinate has doubled. Also, the side lengths have doubled from 3 and 7 units to 6 and 14 units.

This process can be seen in the following picture where the figure on the left represents the original image and the figure on the right represents the dilation:

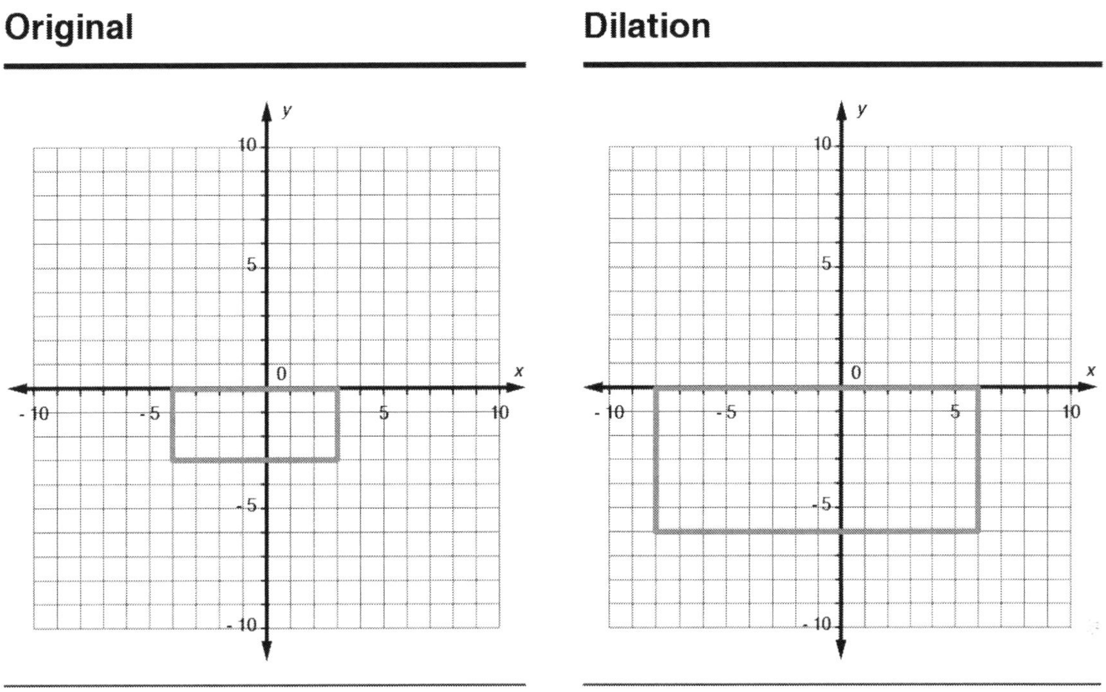


### *Using Algebraic Representations to Explain the Effect of Scale Factors on Dilations*

Consider a shape on a coordinate plane in which any point on the shape can be labeled as $(x,y)$ where $x$ is the $x$-coordinate and $y$ is the $y$-coordinate. If the origin is used as the center of dilation and the scale factor is positive, call it $k$, then a corresponding point on the graph of the dilation is $(kx,ky)$. What that means is that each coordinate in the original figure is just multiplied times the scale factor, resulting in a corresponding coordinate in the new figure of the dilation.

This was the case in the rectangle shown above and is also the case in the octagon shown below:

A Dilation on the Coordinate Plane

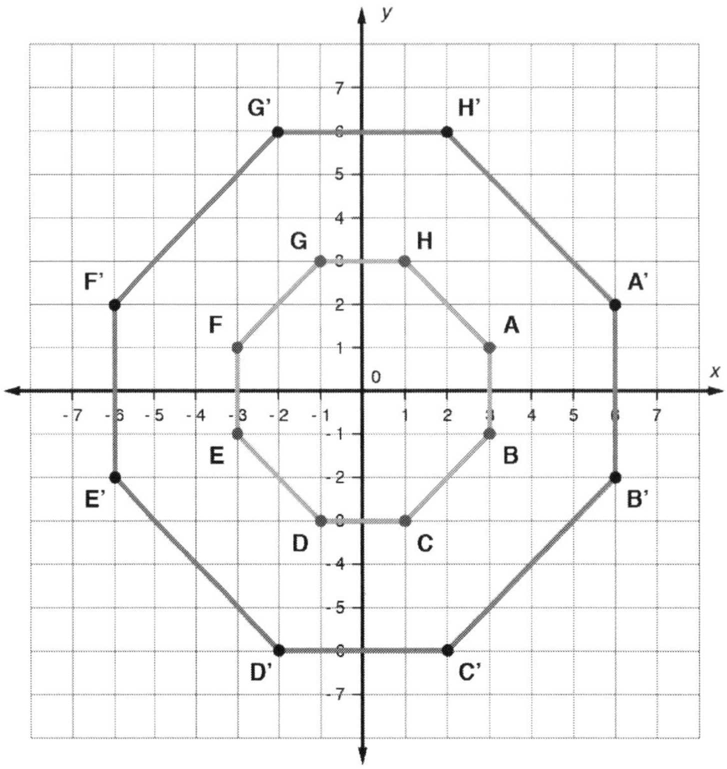


Consider the coordinate (3,1), which represents point A. Its corresponding point in the dilation is A' with coordinates (6,2). Each point has been multiplied times 2. Similarly, G, with ordered pair (-1,3), and G', with ordered pair (-2,6), differ by a multiplicative factor of 2 as well because -1 (the *x*-coordinate of the first ordered pair) times 2 is -2 (the *x*-coordinate of the first ordered pair) and 3 (the *y*-coordinate of the first ordered pair) times 2 is 6 (the *y*-coordinate of the first ordered pair).

## Polar Coordinates

As mentioned, the unit circle has its center at the origin, (0, 0), on the coordinate plane and has a radius of 1. The specific positioning and radius of the unit circle cause its values of *h* and *k* for the standard equation of a circle to equal zero, allowing the unit circle to take on the simplified equation of $x^2 + y^2 = 1$.

Most test takers are more familiar with the rectangular coordinates of $(x, y)$ as points that lie on the square grid of the Cartesian coordinate plane. In this system, the $x$-coordinate represents the number of units in the $x$-direction that the point lies relative to the origin, and the $y$-coordinate represents the number of units in the $y$-direction that the point lies relative to the origin. For example, the coordinate (2, 4) is located two units in the positive $x$ direction and 4 units in the positive $y$ direction from the origin, landing it in the first quadrant.

Instead of using the square grid of the Cartesian coordinate system, the polar coordinate system uses a circular graph. It should be noted that the origin (0, 0) in the polar coordinate system is termed the **pole.** In this system, the locations of points are described using the format $(r, \theta)$, where $r$ represents the radius of the circle or the distance from the pole to the point and $\theta$ is the angle the straight line from the pole to the point makes relative to the positive $x$-axis. Essentially, the coordinates of a point in the polar coordinate system can be considered as the distance the point lies from the pole and the amount it is rotated counterclockwise from the positive $x$-axis.

Returning to the concept of the unit circle, if $\theta$ is the angle going counterclockwise around the origin from the $x$-axis, then the point on the unit circle in that direction will have the coordinates $(\cos\theta, \sin\theta)$.

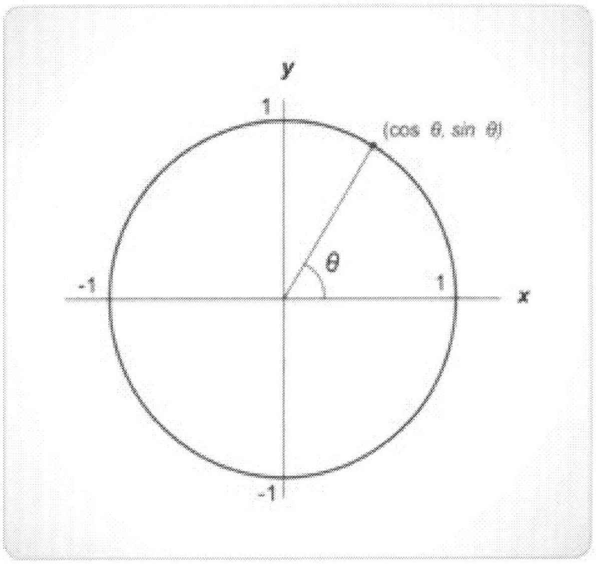


Moreover, using trigonmetic identities, an $x$-coordinate = $r\ cos\theta$ and a $y$-coordinate = $r\ sin\theta$. Therefore:

$$\frac{y}{x} = \frac{r\ cos\theta}{rsin\theta} = tan\theta$$

The inverse or arctan can then be taken of both sides to solve for $\theta$, such that:

$$\theta = tan^{-1}\left(\frac{y}{x}\right)$$

Using the Pythagorean Theorem with the unit circle:

$$r^2 = x^2 + y^2$$

All of this put together helps derive the formula for converting rectangular coordinates $(x, y)$ to polar format $(r, \theta)$:

$$r = \sqrt{x^2 + y^2}, \theta = tan^{-1}\left(\frac{y}{x}\right)$$

when $x \neq 0$.

When $x$ is positive, the positive square root value is used for *r*, and when $x$ is negative, the negative square root value is used for *r*.

Because $x$ is in the denominator, $x$ cannot be equal to zero in this formula. Instead, the following rules apply when $x = 0$:

1. When $y$ is also equal to 0, $\theta = 0$

2. When $y > 0$, $\theta = \frac{\pi}{2}$

3. When $y < 0$, then $\theta = \frac{3\pi}{2}$

The formula used to convert a polar coordinate to rectangular format has essentially already been discussed in reference to the unit circle such that $x = r\cos\theta, y = r\sin\theta$.

# *Three-Dimensional*

## Polygons and Solids

A **polygon** is a closed two-dimensional figure consisting of three or more sides. Polygons can be either convex or concave. A polygon that has interior angles all measuring less than 180° is **convex**. A **concave** polygon has one or more interior angles measuring greater than 180°. Examples are shown below.

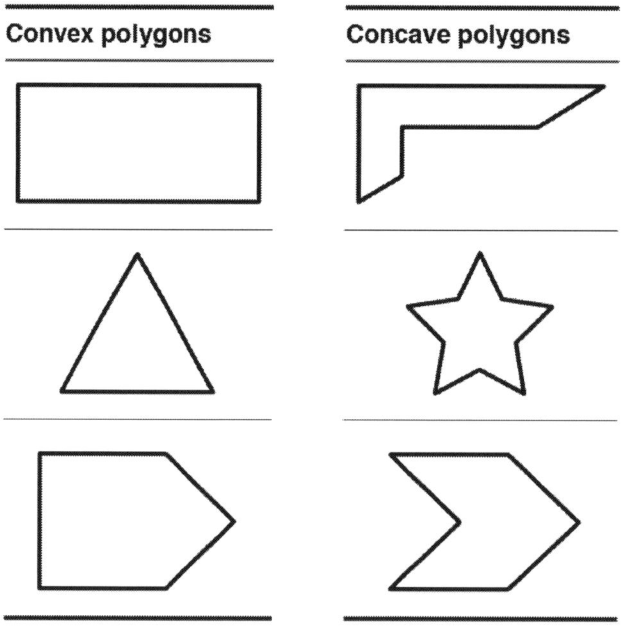

Polygons can be classified by the number of sides (also equal to the number of angles) they have. The following are the names of polygons with a given number of sides or angles:

| # of Sides | Name of Polygon |
|---|---|
| 3 | Triangle |
| 4 | Quadrilateral |
| 5 | Pentagon |
| 6 | Hexagon |
| 7 | Septagon (or heptagon) |
| 8 | Octagon |
| 9 | Nonagon |
| 10 | Decagon |

**Equiangular polygons** are polygons in which the measure of every interior angle is the same. The sides of equilateral polygons are always the same length. If a polygon is both equiangular and equilateral, the polygon is defined as a **regular polygon**. Examples are shown below.

Triangles can be further classified by their sides and angles. A triangle with its largest angle measuring 90° is a right triangle. A triangle with the largest angle less than 90° is an **acute triangle**. A triangle with the largest angle greater than 90° is an **obtuse triangle**.

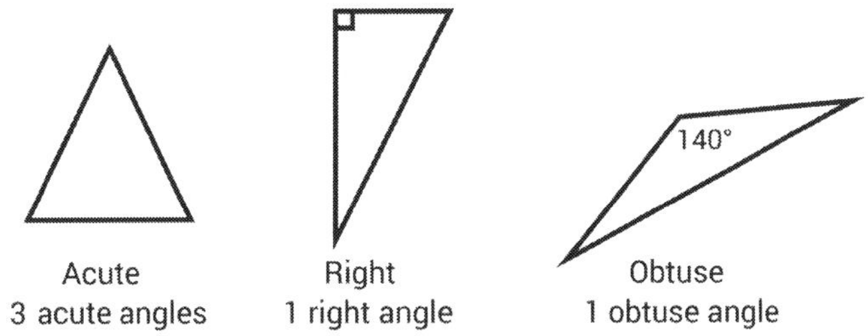

A triangle consisting of two equal sides and two equal angles is an **isosceles triangle**. A triangle with three equal sides and three equal angles is an **equilateral triangle**. A triangle with no equal sides or angles is a **scalene triangle**.

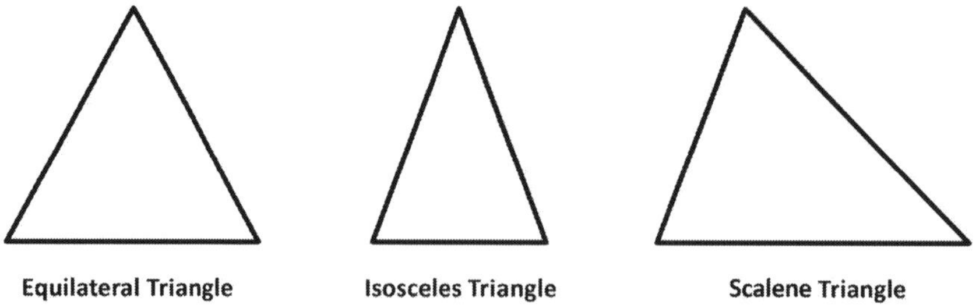


Quadrilaterals can be further classified according to their sides and angles. A quadrilateral with exactly one pair of parallel sides is called a **trapezoid.** A quadrilateral that shows both pairs of opposite sides parallel is a **parallelogram**. Parallelograms include rhombuses, rectangles, and squares. A **rhombus** has four equal sides. A **rectangle** has four equal angles (90° each). A **square** has four 90° angles and four equal sides. Therefore, a square is both a rhombus and a rectangle.

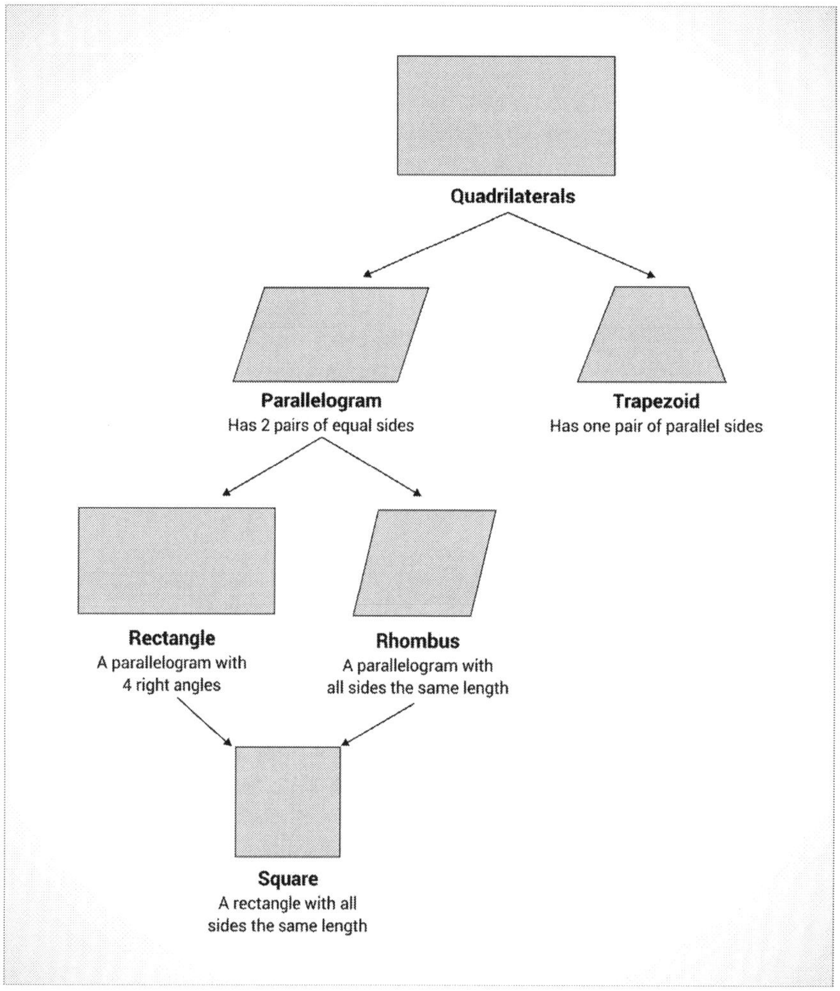

A **solid** is a three-dimensional figure that encloses a part of space. Solids consisting of all flat surfaces that are polygons are called **polyhedrons**. The two-dimensional surfaces that make up a polyhedron are called **faces**. Types of polyhedrons include prisms and pyramids. A **prism** consists of two parallel faces that are **congruent** (or the same shape and same size), and lateral faces going around (which are parallelograms). A prism is further classified by the shape of its base, as shown below:

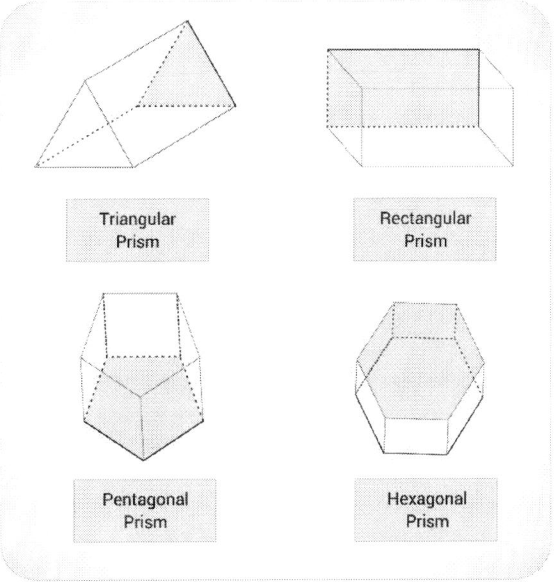


A **pyramid** consists of lateral faces (triangles) that meet at a common point called the **vertex** and one other face that is a polygon, called the **base**. A pyramid can be further classified by the shape of its base, as shown below.

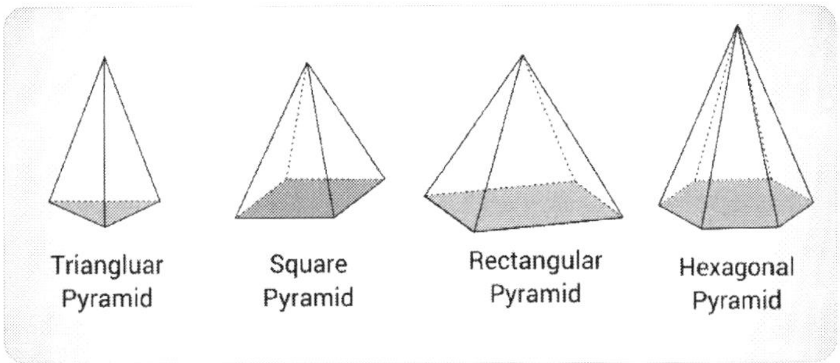


A **tetrahedron** is another name for a triangular pyramid. All the faces of a tetrahedron are triangles.

Solids that are not polyhedrons include spheres, cylinders, and cones. A **sphere** is the set of all points a given distance from a given center point. A sphere is commonly thought of as a three-dimensional circle. A **cylinder** consists of two parallel, congruent (same size) circles and a lateral curved surface. A **cone** consists of a circle as its base and a lateral curved surface that narrows to a point called the **vertex.**

Similar polygons are the same shape but different sizes. More specifically, their corresponding angle measures are congruent (or equal) and the length of their sides is proportional. For example, all sides of one polygon may be double the length of the sides of another. Likewise, similar solids are the same shape but different sizes. Any corresponding faces or bases of similar solids are the same polygons that are proportional by a consistent value.

## Surface Area and Volume

The area of a two-dimensional figure refers to the number of square units needed to cover the interior region of the figure. This concept is similar to wallpaper covering the flat surface of a wall. For example, if a rectangle has an area of 8 square inches (written 8 $in^2$), it will take 8 squares, each with sides one inch in length, to cover the interior region of the rectangle. Note that area is measured in square units such as: square feet or $ft^2$; square yards or $yd^2$; square miles or $mi^2$.

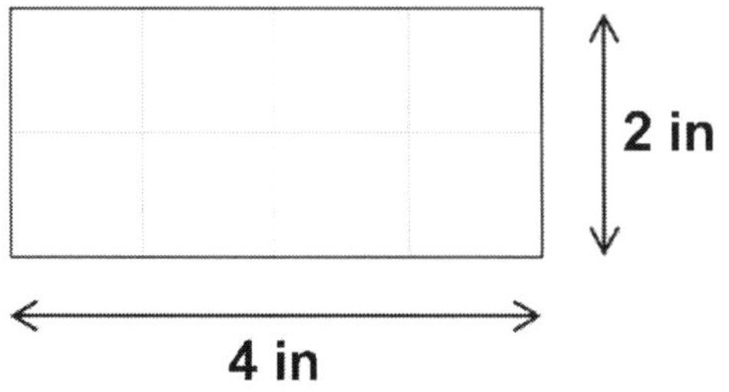


The surface area of a three-dimensional figure refers to the number of square units needed to cover the entire surface of the figure. This concept is similar to using wrapping paper to completely cover the outside of a box. For example, if a triangular pyramid has a surface area of 17 square inches (written $17in^2$), it will take 17 squares, each with sides one inch in length, to cover the entire surface of the pyramid. Surface area is also measured in square units.

Many three-dimensional figures (solid figures) can be represented by nets consisting of rectangles and triangles. The surface area of such solids can be determined by adding the areas of each of its faces and bases. Finding the surface area using this method requires calculating the areas of rectangles and triangles. To find the area (*A*) of a rectangle, the length (*l*) is multiplied by the width $(w) \rightarrow A = l \times w$. The area of a rectangle with a length of 8cm and a width of 4cm is calculated:

$$A = (8cm) \times (4cm) \rightarrow A = 32cm^2$$

To calculate the area (*A*) of a triangle, the product of $\frac{1}{2}$, the base (*b*), and the height (*h*) is found:

$$A = \frac{1}{2} \times b \times h$$

Note that the height of a triangle is measured from the base to the vertex opposite of it forming a right angle with the base. The area of a triangle with a base of 11cm and a height of 6cm is calculated:

$$A = \frac{1}{2} \times (11cm) \times (6cm) \rightarrow A = 33cm^2$$

Consider the following triangular prism, which is represented by a net consisting of two triangles and three rectangles.

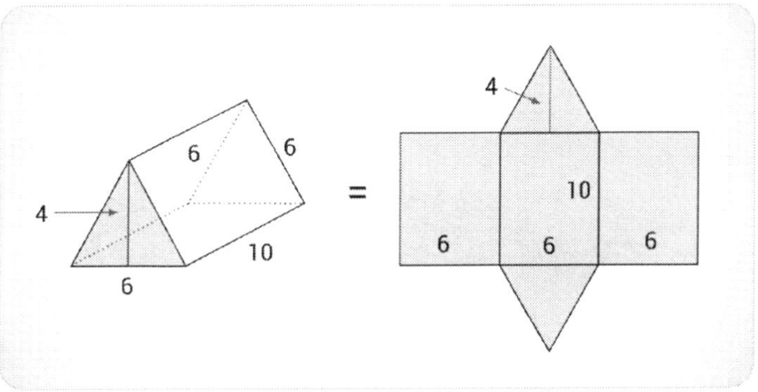


The surface area of the prism can be determined by adding the areas of each of its faces and bases. The surface area (*SA*) = area of triangle + area of triangle + area of rectangle + area of rectangle + area of rectangle.

$$SA = \left(\frac{1}{2} \times b \times h\right) + \left(\frac{1}{2} \times b \times h\right) + (l \times w) + (l \times w) + (l \times w)$$

$$SA = \left(\frac{1}{2} \times 6 \times 4\right) + \left(\frac{1}{2} \times 6 \times 4\right) + (6 \times 10) + (6 \times 10) + (6 \times 10)$$

$$SA = (12) + (12) + (60) + (60) + (60)$$

$$SA = 204 \textit{ square units}$$

**Volume** is the measurement of how much space an object occupies, like how much space is in the cube. Volume is useful in determining the space within a certain three-dimensional object. Volume can be calculated for a cube, rectangular prism, cylinder, pyramid, cone, and sphere. By knowing specific dimensions of the objects, the volume of the object is computed with these figures. The units for the volumes of solids can include cubic centimeters, cubic meters, cubic inches, and cubic feet.

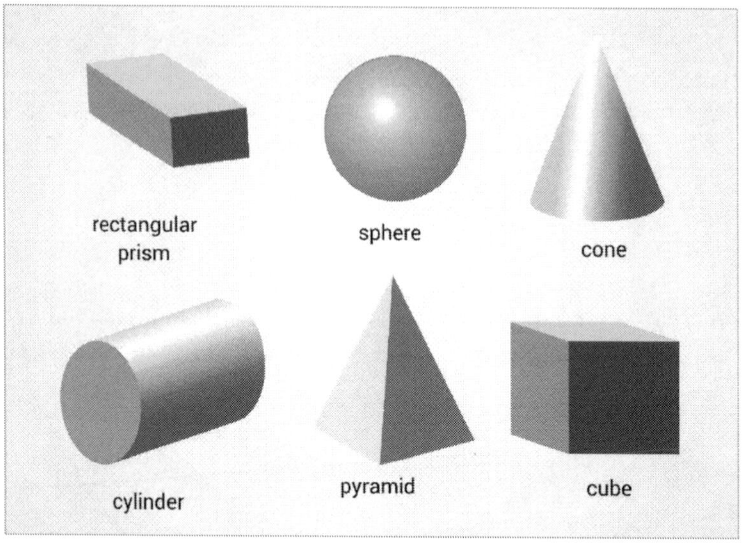


### *Cube*

The cube is the simplest figure for which volume can be determined because all dimensions in a cube are equal. In the following figure, the length, width, and height of the cube are all represented by the variable *a* because these measurements are equal lengths.

The volume of any rectangular, three-dimensional object is found by multiplying its length by its width by its height. In the case of a cube, the length, width, and height are all equal lengths, represented by the variable *a*. Therefore, the equation used to calculate the volume is $(a \times a \times a)$ or $a^3$. In a real-world example of this situation, if the length of a side of the cube toy block is 3 centimeters, the volume is calculated by utilizing the formula $(3 \times 3 \times 3) = 9$ $cm^3$.

### *Rectangular Prism*

The dimensions of a rectangular prism are not necessarily equal as those of a cube. Therefore, the formula for a rectangular prism recognizes that the dimensions vary and use different variables to represent these lengths. The length, width, and height of a rectangular prism can be represented with the variables *a, b*, and *c*.

The equation used to calculate volume is length times width times height. Using the variables in the diagram above, this means $a \times b \times c$. In a real-world application of this situation, if $a$=2 cm, $b$=3 cm, and $c$=4 cm, the volume is calculated by utilizing the formula $3 \times 4 \times 5 = 60$ $cm^3$.

### *Cylinder*

Discovering a cylinder's volume requires the measurement of the cylinder's base, length of the radius, and height. The height of the cylinder can be represented with variable *h*, and the radius can be represented with variable *r*.

The formula to find the volume of a cylinder is $\pi r^2 h$. Notice that $\pi r^2$ is the formula for the area of a circle. This is because the base of the cylinder is a circle. To calculate the volume of a cylinder, the slices of circles needed to build the entire height of the cylinder are added together. For example, if the radius is 5 feet and the height of the cylinder is 10 feet, the cylinder's volume is calculated by using the following equation: $\pi 5^2 \times 10$. Substituting 3.14 for $\pi$, the volume is 785.4 $ft^3$.

### *Pyramid*

To calculate the volume of a pyramid, the area of the base of the pyramid is multiplied by the pyramid's height by $\frac{1}{3}$. The area of the base of the pyramid is found by multiplying the base length by the base width.

Therefore, the formula to calculate a pyramid's volume is $(L \times W \times H) \div 3$.

### *Cone*

The formula to calculate the volume of a circular cone is similar to the formula for the volume of a pyramid. The primary difference in determining the area of a cone is that a circle serves as the base of a cone. Therefore, the area of a circle is used for the cone's base.

The variable *r* represents the radius, and the variable *h* represents the height of the cone. The formula used to calculate the volume of a cone is $\frac{1}{3}\pi r^2 h$. Essentially, the area of the base of the cone is multiplied by the cone's height. In a real-life example where the radius of a cone is 2 meters and the height of a cone is 5 meters, the volume of the cone is calculated by utilizing the formula:

$$\frac{1}{3}\pi 2^2 \times 5 = 21$$

After substituting 3.14 for $\pi$, the volume is 785.4 $ft^3$.

### *Sphere*

The volume of a sphere uses $\pi$ due to its circular shape.

The length of the radius, *r*, is the only variable needed to determine the sphere's volume. The formula to calculate the volume of a sphere is $\frac{4}{3}\pi r^3$. Therefore, if the radius of a sphere is 8 centimeters, the volume of the sphere is calculated by utilizing the formula:

$$\frac{4}{3}\pi(8)^3 = 2{,}143\ cm^3$$

## Three-Dimensional Figures with Nets

A **net** is a construction of two-dimensional figures that can be folded to form a given three-dimensional figure. More than one net may exist to fold and produce the same solid, or three-dimensional figure. The bases and faces of the solid figure are analyzed to determine the polygons (two-dimensional figures) needed to form the net.

Consider the following triangular prism:

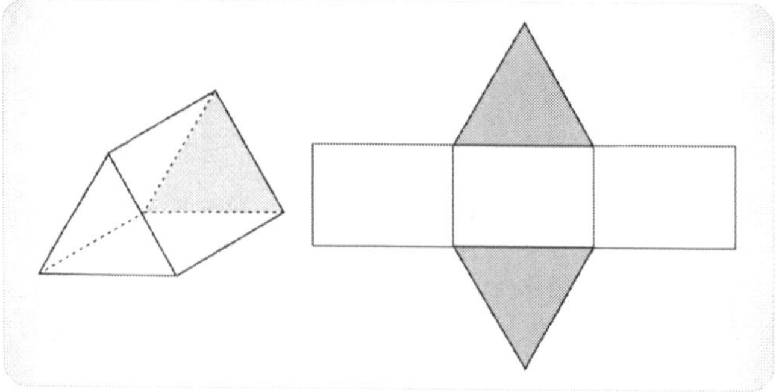

The surface of the prism consists of two triangular bases and three rectangular faces. The net beside it can be used to construct the triangular prism by first folding the triangles up to be parallel to each other, and then folding the two outside rectangles up and to the center with the outer edges touching.

Consider the following cylinder:

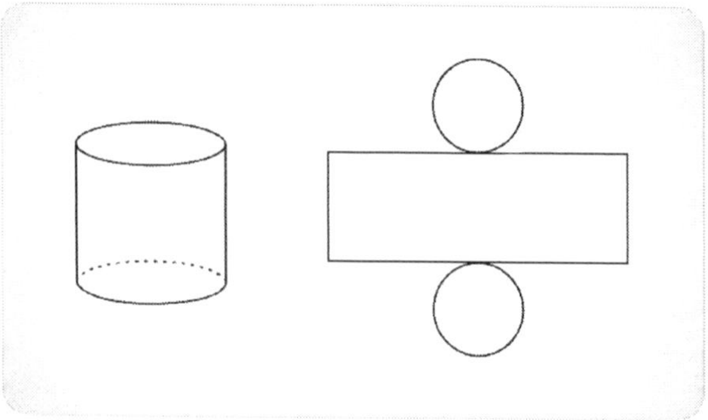

The surface consists of two circular bases and a curved lateral surface that can be opened and flattened into a rectangle. The net beside it can be used to construct the cylinder by first folding the circles up to be parallel to each other, and then curving the sides of the rectangle up to touch each other. The top and bottom of the folded rectangle should be touching the outside of both circles.

Consider the following square pyramid below on the left. The surface consists of one square base and four triangular faces. The net below on the right can be used to construct the square pyramid by folding each triangle towards the center of the square. The top points of the triangle meet at the vertex.

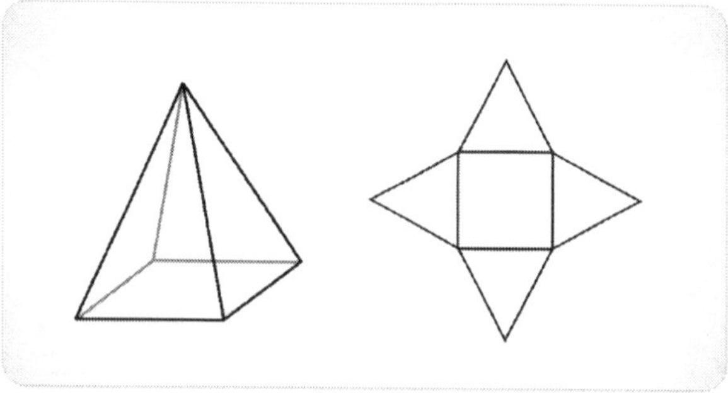

### Simplifying Three-Dimensional Objects

Three-dimensional objects can be simplified into related two-dimensional shapes to solve problems. This simplification can make problem-solving a much easier experience. An isometric representation of a three-dimensional object can be completed so that important properties (e.g., shape, relationships of faces and surfaces) are noted. Edges and vertices can be translated into two-dimensional objects as well. The net represents the sum of the three different faces. Depending on the problem, using a smaller portion of the given shape may be helpful, by simplifying the steps necessary to solve.

## Coordinates in Three Dimensions

The graph of a function of one variable can be represented in the $xy$-plane and is known as a **curve.** When a function has two variables, the function is graphed in three-dimensional space, and the graph is known as a **surface.** The graph is the set of all ordered triples (*x, y, z*) that satisfy the function. Within three-dimensional space, there is a third axis known as the ***z*-axis**. The following graphic labels points using all three axes:

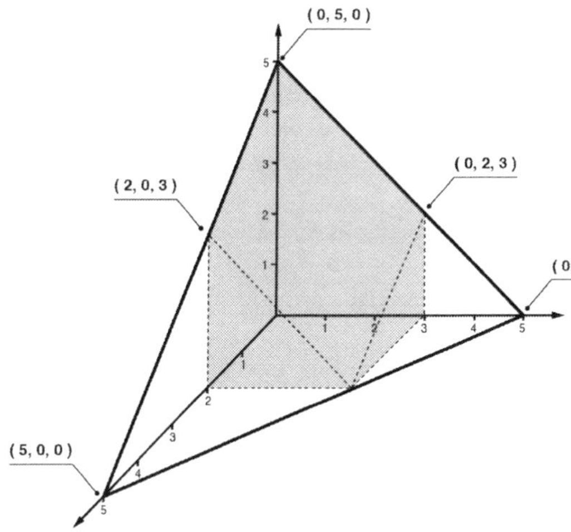


# *Trigonometry*

## Trigonometric Ratios in Right Triangles

### Pythagorean Theorem

The Pythagorean theorem is an important tenet in geometry. It states that for right triangles, the sum of the squares of the two shorter sides will be equal to the square of the longest side (also called the **hypotenuse**). The longest side will always be the side opposite to the 90° angle. If this side is called *c*, and the other two sides are *a* and *b*, then the Pythagorean theorem states that:

$$c^2 = a^2 + b^2$$

Since lengths are always positive, this also can be written as:

$$c = \sqrt{a^2 + b^2}$$

A diagram to show the parts of a triangle using the Pythagorean theorem is below.

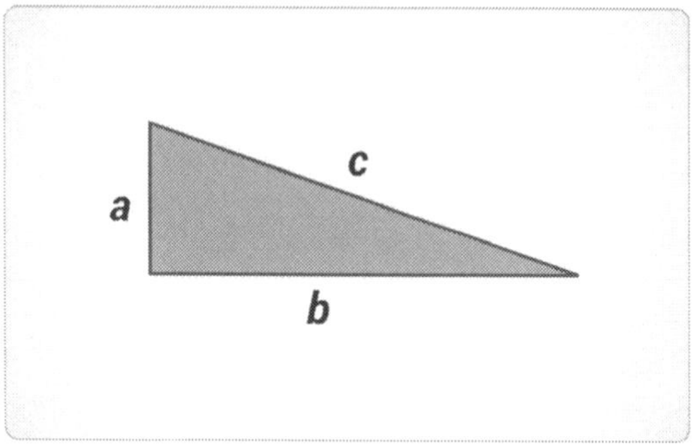


As an example of the theorem, suppose that Shirley has a rectangular field that is 5 feet wide and 12 feet long, and she wants to split it in half using a fence that goes from one corner to the opposite corner. How long will this fence need to be? To figure this out, note that this makes the field into two right triangles, whose hypotenuse will be the fence dividing it in half. Therefore, the fence length will be given by:

$$\sqrt{5^2 + 12^2} = \sqrt{169} = 13 \text{ feet long}$$

## Trigonometric Functions

Within similar triangles, corresponding sides are proportional, and angles are congruent. In addition, within similar triangles, the ratio of the side lengths is the same. This property is true even if side lengths are different. Within right triangles, trigonometric ratios can be defined for the acute angle within the triangle. The functions are defined through ratios in a right triangle. Sine of acute angle, A, is opposite over hypotenuse, cosine is adjacent over hypotenuse, and tangent is opposite over adjacent. Note that expanding or shrinking the triangle won't change the ratios. However, changing the angle measurements will alter the calculations.

## Complementary Angles

Angles that add up to 90 degrees are **complementary**. Within a right triangle, two complementary angles exist because the third angle is always 90 degrees. In this scenario, the sine of one of the complementary angles is equal to the cosine of the other angle. The opposite is also true. This relationship exists because sine and cosine will be calculated as the ratios of the same side lengths.

## Application of Trigonometry to General Triangles

### *The Area Formula*

A triangle that isn't a right triangle is known as an **oblique triangle**. It should be noted that even if the triangle consists of three acute angles, it is still referred to as an oblique triangle. *Oblique*, in this case, does not refer to an angle measurement. Consider the following oblique triangle:

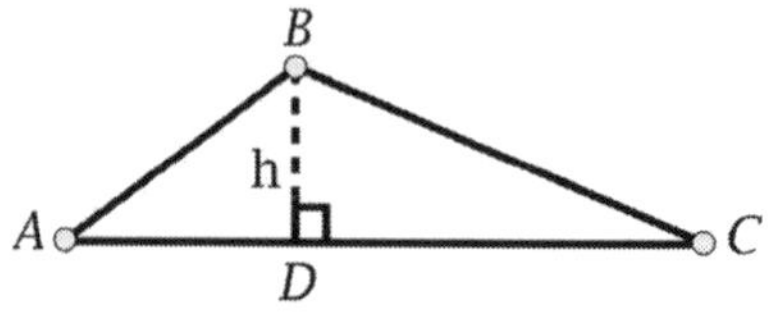

For this triangle:

$$Area = {}^{1}/_{2} \times base \times height = {}^{1}/_{2} \times AC \times BD$$

The auxiliary line drawn from the vertex B perpendicular to the opposite side AC represents the height of the triangle. This line splits the larger triangle into two smaller right triangles, which allows for the use of the trigonometric functions (specifically that $\sin A = {}^{h}/_{AB}$). Therefore:

$$Area = {}^{1}/_{2} \times AC \times AB \times \sin A$$

Typically the sides are labelled as the lowercase letter of the vertex that's opposite. Therefore, the formula can be written as $Area = \frac{1}{2}ab \sin A$. This area formula can be used to find areas of triangles when given side lengths and angle measurements, or it can be used to find side lengths or angle measurements based on a specific area and other characteristics of the triangle.

### *Medians, Midpoints, and Altitudes*

A **median** of a triangle is the line drawn from a vertex to the midpoint on the opposite side. A triangle has three medians, and their point of intersection is known as the **centroid.** An **altitude** is a line drawn from a vertex perpendicular to the opposite side. A triangle has three altitudes, and their point of intersection is known as the **orthocenter**. An altitude can actually exist outside, inside, or on the triangle depending on the placement of the vertex. Many problems involve these definitions. For example, given one endpoint of a line segment and the midpoint, the other endpoint can be determined by using the midpoint formula. In addition, area problems heavily depend on these definitions. For example, it can be proven that the median of a triangle divides it into two regions of equal areas. The actual formula for the area of a triangle depends on its altitude.

### *Special Triangles*

An **isosceles triangle** contains at least two equal sides. Therefore, it must also contain two equal angles and, subsequently, contain two medians of the same length. An isosceles triangle can also be labelled as an **equilateral triangle** (which contains three equal sides and three equal angles) when it meets these conditions. In an equilateral triangle, the measure of each angle is always 60 degrees. Also within an equilateral triangle, the medians are of the same length. A **scalene triangle** can never be an equilateral or an isosceles triangle because it contains no equal sides and no equal angles. Also, medians in a scalene triangle can't have the same length. However, a **right triangle**, which is a triangle containing a 90-degree angle, can be a scalene triangle. There are two types of special right triangles. The **30-60-90 right triangle** has angle measurements of 30 degrees, 60 degrees, and 90 degrees. Because of the nature of this triangle, and through the use of the Pythagorean theorem, the side lengths have a special relationship. If *x* is the length opposite the 30-degree angle, the length opposite the 60-degree angle is $\sqrt{3}x$, and the hypotenuse has length $2x$. The **45-45-90 right triangle** is also special as it contains two angle measurements of 45 degrees. It can be proven that, if *x* is the length of the two equal sides, the hypotenuse is $x\sqrt{2}$. The properties of all of these special triangles are extremely useful in determining both side lengths and angle measurements in problems where some of these quantities are given and some are not.

## Trigonometric Identities

From the unit circle, the trigonometric ratios were found for the special right triangle with a hypotenuse of 1.

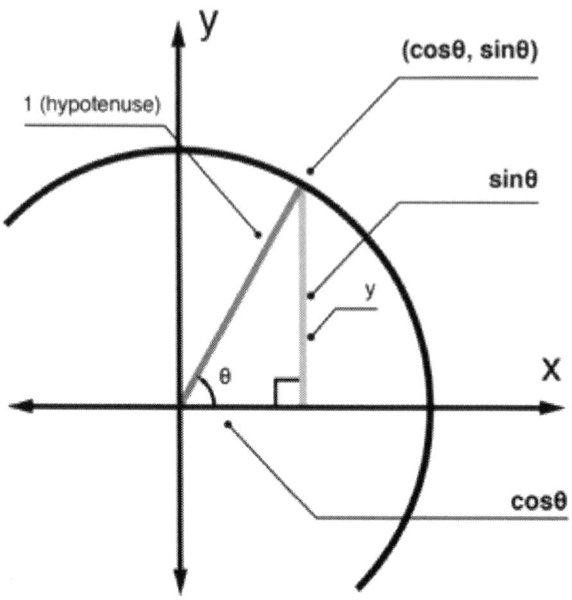


From this triangle, the following Pythagorean identities are formed:

$$\sin^2\theta + \cos^2\theta = 1$$

$$\tan^2\theta + 1 = \sec^2\theta$$

$$1 + \cot^2\theta = \csc^2\theta$$

The second two identities are formed by manipulating the first identity. Since identities are statements that are true for any value of the variable, then they may be used to manipulate equations. For example, a problem may ask for simplification of the expression:

$$\cos^2 x + \cos^2 x \tan^2 x$$

Using the fact that $\tan(x) = \frac{\sin x}{\cos x}, \frac{\sin^2 x}{\cos^2 x}$ can then be substituted in for $\tan^2 x$, making the expression:

$$\cos^2 x + \cos^2 x \frac{\sin^2 x}{\cos^2 x}$$

Then the two $\cos^2 x$ terms on top and bottom cancel each other out, simplifying the expression to $\cos^2 x + \sin^2 x$. By the first Pythagorean identity stated above, the expression can be turned into:

$$\cos^2 x + \sin^2 x = 1$$

Other trigonometric identities such as half-angle formulas, sum and difference formulas, and difference of angles formulas can be used to prove and rewrite trigonometric equations. Depending on the given equation or expression, the correct identities need to be chosen to write equivalent statements.

The graph of sine is equal to the graph of cosine, shifted $\frac{\pi}{2}$ units. Therefore, the function $y = \sin x$ is equal to:

$$y = \cos(\frac{\pi}{2} - x)$$

Within functions, adding a constant to the independent variable shifts the graph either left or right. By shifting the cosine graph, the curve lies on top of the sine function. By transforming the function, the two equations give the same output for any given input.

## Radian Measure

**Degrees** are used to express the size of an angle. A complete circle is represented by 360°, and a half circle is represented by 180°. In addition, a right angle fills one quarter of a circle and is represented by 90°.

**Radians** are another way to denote angles in terms of $\pi$, rather than degrees. A complete circle is represented by $2\pi$ radians. The formula used to convert degrees to radians is:

$$Radians = \frac{degrees \times \pi}{180}$$

For example, to convert 270 degrees to radians:

$$Radians = \frac{270 \times \pi}{180} = 4.71$$

As previously mentioned, the **arc of a circle** is the distance between two points on the circle. The length of the arc of a circle in terms of degrees is easily determined if the value of the central angle is known. The length of the arc is simply the value of the central angle. In this example, the length of the arc of the circle in degrees is 75°.

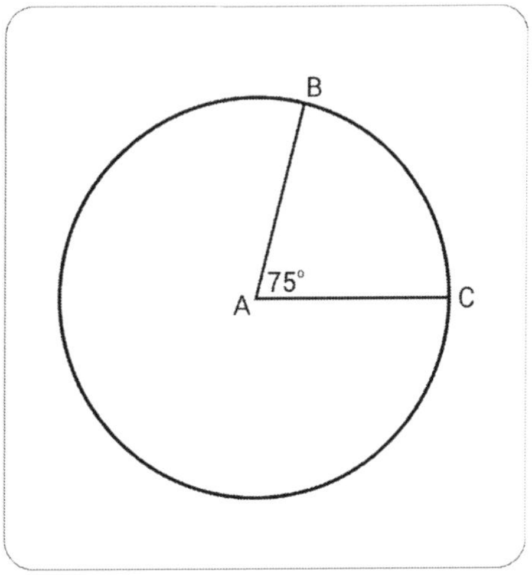

To determine the length of the arc of a circle in distance, the values for both the central angle and the radius must be known. This formula is:

$$\frac{central\ angle}{360°} = \frac{arc\ length}{2\pi r}$$

The equation is simplified by cross-multiplying to solve for the arc length.

In the following example, the student should substitute the values of the central angle (75°) and the radius (10 inches) into the equation above to solve for the arc length.

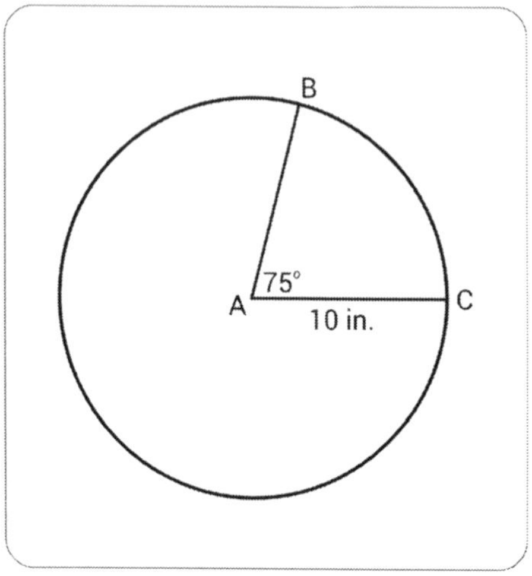


$$\frac{75°}{360°} = \frac{\text{arc length}}{2(3.14)(10\text{in.})}$$

To solve the equation, first cross-multiply: 4710 = 360(arc length). Next, divide each side of the equation by 360. The result of the formula is that the arc length is 13.1 inches (rounded). Please note that arc length is often referred to as ***s***.

As a special technological note for trigonometric functions, when finding the trigonometric function or an angle on the calculator, make a note using degrees or radians to get the correct value. Whether computing the sine of $\frac{\pi}{6}$ or computing the sine of 30°, the answer should come out to $\frac{1}{2}$. However, there is usually a "Mode" function on the calculator to select either radian or degree.

## Laws of Sines and Cosines

The **law of sines** and **law of cosines** are two more relationships that exist within oblique triangles. Consider a triangle with sides *a*, *b*, and *c*, and angles *A*, *B*, and *C* opposite the corresponding sides.

The law of cosines states that:

$$c^2 = a^2 + b^2 - 2ab\ \cos C$$

The law of sines states that:

$$\frac{\sin A}{a} = \frac{\sin B}{b} = \frac{\sin C}{c}$$

In addition to the area formula, these two relationships can help find unknown angle and side measurements in oblique triangles.

## Equations

Two important laws to know are the Law of Cosines and the Law of Sines. With memorization, they allow one to quickly solve some trigonometry problems involving non-right triangles. Consider the following triangle *(a)* with sides *a, b,* and *c,* and angles *A, B,* and *C*:

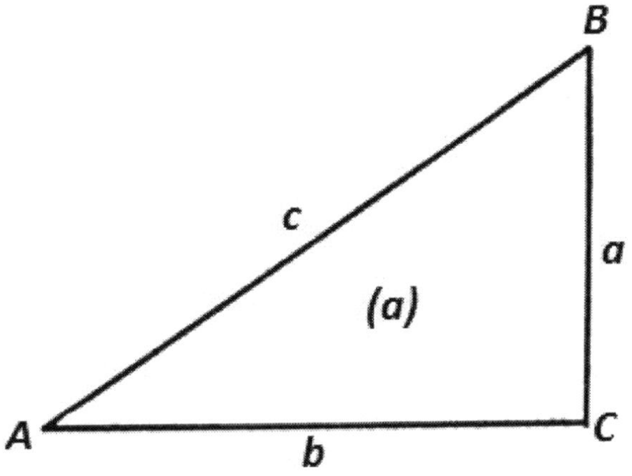


The Law of Sines states that $\frac{a}{\sin A} = \frac{b}{\sin B} = \frac{c}{\sin C}$ and the Law of Cosines states that:

$$c^2 = a^2 + b^2 - 2ab\cos C$$

The Law of Cosines is helpful when two sides and the included angle are known, and the third side needs to be found (SAS). Also, it is helpful when three sides are known, and any angle needs to be found (SSS). For example, consider finding the missing side *x* in the following triangle where the angles are given in degrees:

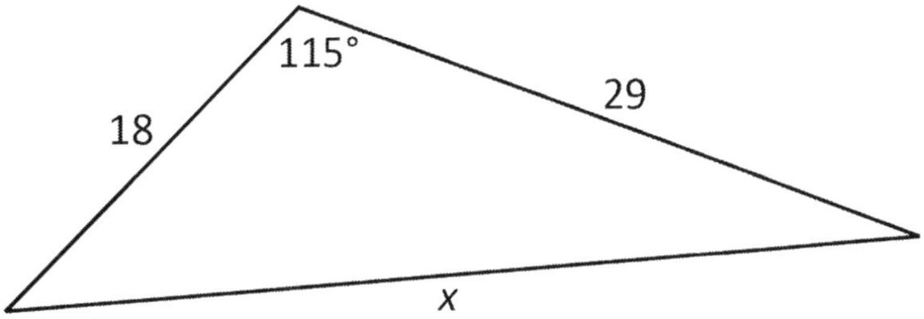


The Law of Cosines would give: $x^2 = 18^2 + 29^2 - 2 \times 18 \times 29 \times \cos(115°) = 1606.21$, rounded to the nearest hundredth. Therefore, taking the positive square root results in $x = 40.08$.

The Law of Sines is helpful when two angles and one side of the triangles are known (AAS or ASA) or two sides and an angle opposite are known (SSA). For example, consider finding the missing side *b* in the following triangle:

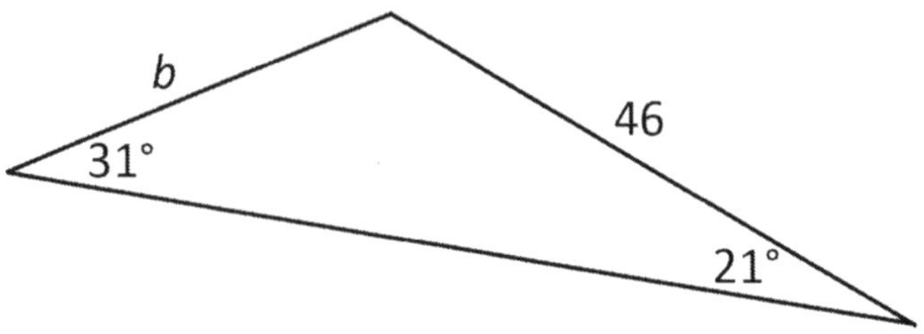


The Law of Sines states that:

$$\frac{46}{\sin 31} = \frac{b}{\sin 21}$$

By finding cross products and dividing both sides of the equation by $\sin 31$, we have:

$$b = \frac{46 \sin 21}{\sin 31} = 32 \text{ m}$$

Knowing how to solve trigonometry equations is also important. It is crucial to know how to use the unit circle (as shown before) in these instances.

## The Unit Circle

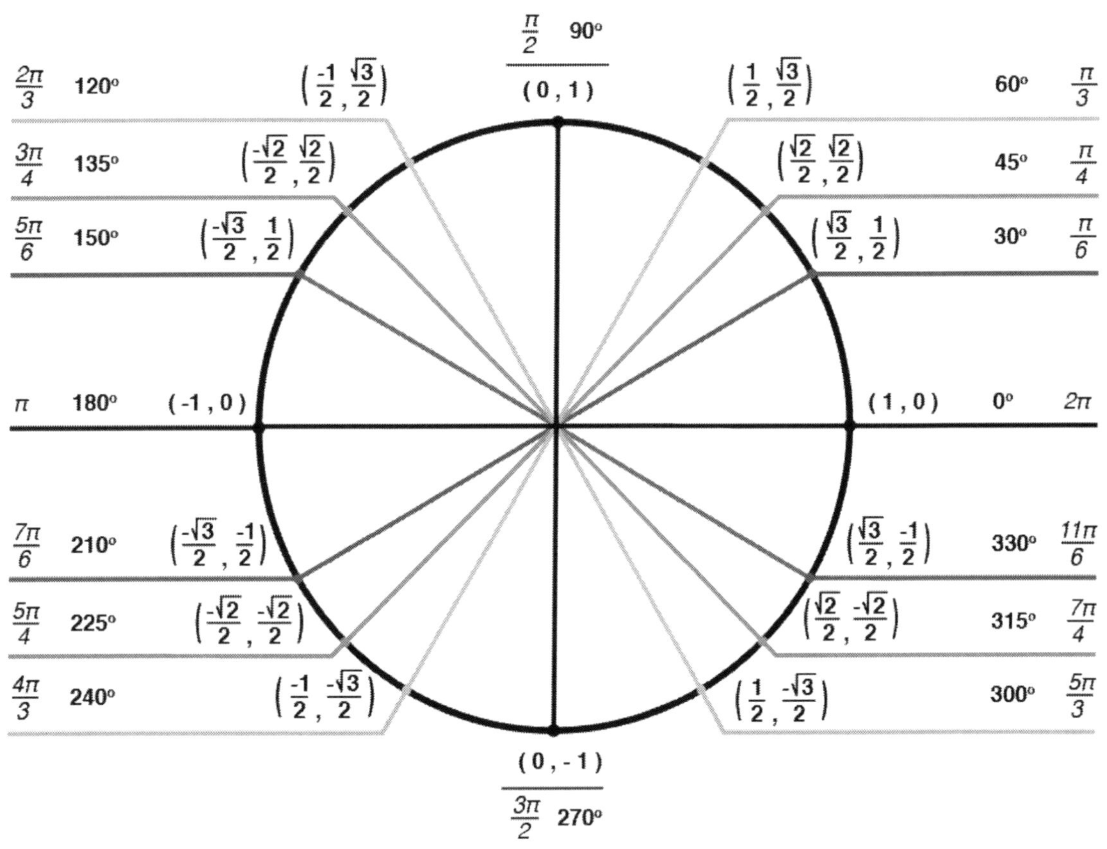


A simple example would be to solve $3\cos x + 1 = -5$ on the interval $0 \leq x \leq 2\pi$. Solving for $\cos x$ results in the equation $\cos x = -1/2$. One would then look at the unit circle to find the angle(s) that correspond to an *x*-coordinate of $-1/2$. There are two solutions:

$$x = \frac{2\pi}{3} \text{ and } x = \frac{4\pi}{3}$$

A more complicated example includes solving the equation $\cos^2(\theta) - \cos(\theta) - 2 = 0$ on the interval $0 \leq x \leq 2\pi$. This has a similar form as a quadratic equation, and it can be factored into:

$$(\cos(\theta) - 2)(\cos(\theta) + 1) = 0$$

Each factor is then set equal to zero and the unit circle can be used to solve for $\theta$. The first factor results in the equation $\cos(\theta) = 2$, which has no solution because the range of $\cos(\theta)$ is $[-1,1]$. The second factor results in the equation $\cos(\theta) = -1$, which has solution $\theta = \pi$. Therefore, this is the only solution to the original equation.

The previous example shows why it is important to know the range of the trigonometric functions. Consider the following equation to be solved:

$$\tan^2\theta + 1 = 0$$

In order to solve this, 1 is subtracted from both sides, resulting in:

$$\tan^2\theta = -1$$

At first, one might try to take the square root of both sides. However, this equation has no solution because the square of any function cannot be a negative number. It is important to watch out for such situations.

Some problems might also incorporate applying trigonometric identities. For example, consider solving the equation $\cos x + \sin x = 1$ on the interval:

$$0 \le x < 2\pi$$

At first glance, it looks like nothing can be done to isolate the variable *x*. However, squaring both sides results in:

$$\cos^2 x + 2\sin x\cos x + \sin^2 x = 1$$

We know that $\sin^2 x + \cos^2 x = 1$ from the Pythagorean Identity, so this equation simplifies to $1 + 2\sin x\cos x = 1$, or $2\sin x\cos x = 0$. Therefore, we need to solve $\sin x = 0$ and $\cos x = 0$ on the interval:

$$0 \le x < 2\pi$$

This results in the following solutions: $x = 0, \frac{\pi}{2}, \pi, \frac{3\pi}{2}$. Now, because we squared the original problem, we need to check these solutions into the original equation to see if they are true solutions. Plugging $\frac{\pi}{2}$ and $\frac{3\pi}{2}$ do not work. Therefore, the only solutions to the original problem are $x = 0$ and $\pi$.

## Double-Angle Formulas

Another set of trigonometric identities are the double-angle formulas:

$$\sin 2\alpha = 2\sin\alpha\,\cos\alpha$$

$$\cos 2\alpha = \begin{cases} \cos^2\alpha - \sin^2\alpha \\ 2\cos^2\alpha - 1 \\ 1 - 2\sin^2\alpha \end{cases}$$

Using these formulas, the following identity can be proved:

$$\sin 2x = \frac{2\tan x}{1 + \tan^2 x}$$

By using one of the Pythagorean identities, the denominator can be rewritten as $1 + \tan^2 x = \sec^2 x$. By knowing the reciprocals of the trigonometric identities, the secant term can be rewritten to form the equation:

$$\sin 2x = \frac{2 \tan x}{1} \times \cos^2 x$$

Replacing $\tan(x)$, the equation becomes $\sin 2x = \frac{2 \sin x}{\cos x} \times \cos^2 x$, where the $\cos x$ can cancel out. The new equation is:

$$\sin 2x = 2 \sin x * \cos x$$

This final equation is one of the double-angle formulas.

## Practice Questions

1. Which of the following shows a line of symmetry?

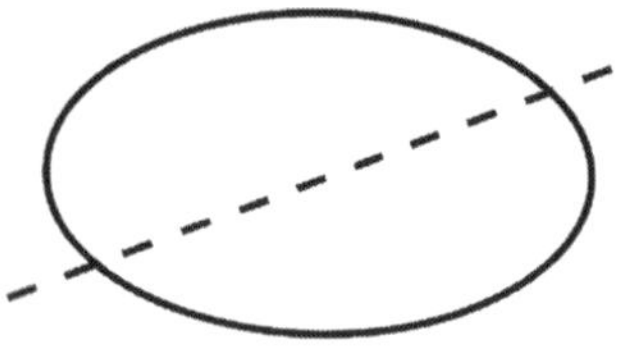

a.

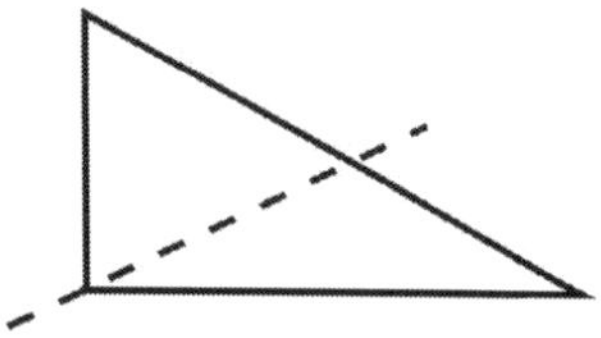

b.

c.

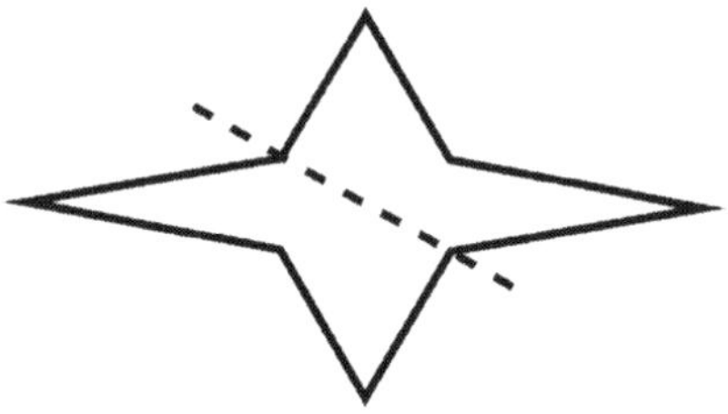

d.

e. They all show a line of symmetry.

2. A line goes through the point (-4, 0) and the point (0, 2). What is the slope of the line?

a. 2

b. 4

c. $\frac{3}{2}$

d. $\frac{1}{2}$

e. -2

3. In the *xy*-plane, the graph of $y = x^2 + 2$ and the circle with center $(0,1)$ and radius 1 have how many points of intersection?

a. 0
b. 1
c. 2
d. 3
e. 4

4. What is the volume of a rectangular prism with the height of 3 centimeters, a width of 5 centimeters, and a depth of 11 centimeters?

a. 19 $cm^3$
b. 165 $cm^3$
c. 225 $cm^3$
d. 150 $cm^3$
e. 88 $cm^3$

5. What is the volume of a pyramid, with the area of the base measuring 12 $inches^2$, and the height measuring 15 inches?

a. 180 $in^3$
b. 90 $in^3$
c. 30 $in^3$
d. 60 $in^3$
e. 45 $in^3$

6. A pizzeria owner regularly creates jumbo pizzas, each with a radius of 9 inches. She is mathematically inclined, and wants to know the area of the pizza to purchase the correct boxes and know how much she is feeding her customers. What is the area of the circle, in terms of $\pi$, with a radius of 9 inches?

a. $3\pi$ $in^2$
b. $18\pi$ $in^2$
c. $90\pi$ $in^2$
d. $9\pi$ $in^2$
e. $81\pi$ $in^2$

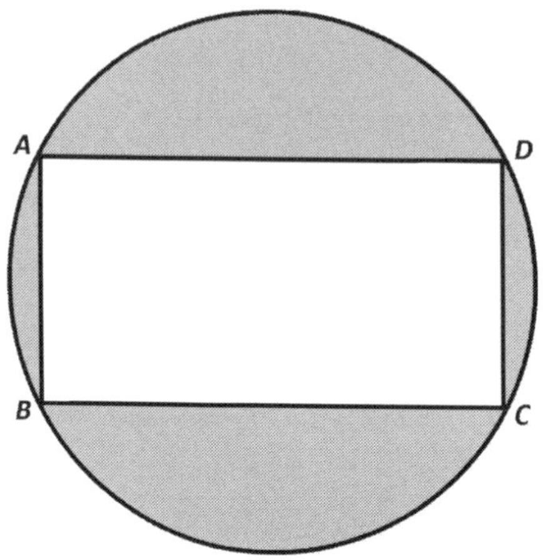


7. Rectangle *ABCD* is inscribed in the circle above. The length of side *AB* is 9 inches and the length of side *BC* is 12 inches. What is the area of the shaded region?

a. 64.4 sq. in.
b. 68.6 sq. in.
c. 62.8 sq. in.
d. 61.3 sq. in.
e. 64.6 sq. in.

8. For the following similar triangles, what are the values of $x$ and $y$ (rounded to one decimal place)?

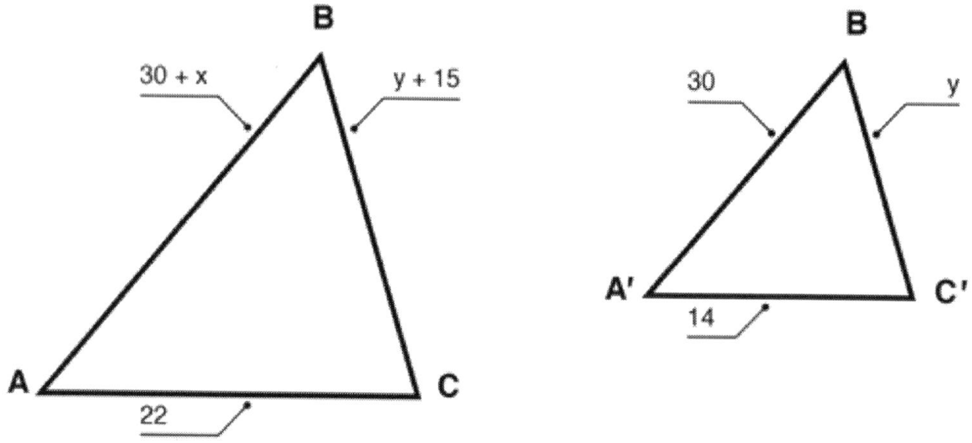


a. $x = 16.5, y = 25.1$
b. $x = 19.5, y = 24.1$
c. $x = 17.1, y = 26.3$
d. $x = 26.3, y = 17.1$
e. $x = 24.1, y = 19.5$

9. If the volume of a sphere is $288\pi$ cubic meters, what are the radius and surface area of the same sphere?

a. Radius 6 meters and surface area $144\pi$ square meters
b. Radius 36 meters and surface area $144\pi$ square meters
c. Radius 6 meters and surface area $12\pi$ square meters
d. Radius 36 meters and surface area $12\pi$ square meters
e. Radius 12 meters and surface area $144\pi$ square meters

10. The triangle shown below is a right triangle. What's the value of $x$?

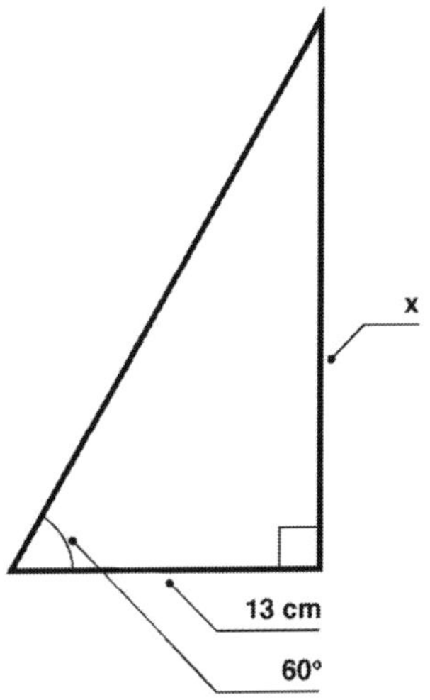


a. $x = 1.73$
b. $x = 0.57$
c. $x = 13$
d. $x = 14.73$
e. $x = 22.52$

11. Using the following diagram, calculate the total circumference, rounding to the nearest decimal place:

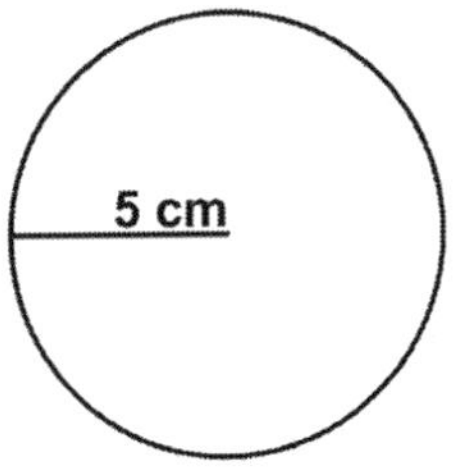


a. 25.0 cm
b. 15.7 cm
c. 78.5 cm
d. 50.0 cm
e. 31.4 cm

12. An angle measures 54 degrees. In order to correctly determine the measure of its complementary angle, what concept is necessary?

a. Two complementary angles sum up to 180 degrees.
b. Complementary angles are always acute.
c. Two complementary angles sum up to 90 degrees.
d. Complementary angles sum up to 360 degrees.
e. At least one of the angles of two complementary angles must be obtuse.

13. A triangular pyramid is composed of all equilateral triangles with sides measuring 8 cm. What is the surface area of the pyramid?

a. 128 sq. cm.
b. 110.7 sq. cm.
c. 114.6 sq. cm.
d. 117.8 sq. cm.
e. 27.68 sq. cm.

14. What are the center and radius of a circle with equation $2x^2 + 2y^2 - 16x - 8y + 32 = 0$?

a. Center (2, 4) and radius 2
b. Center (4, 2) and radius 2
c. Center (4, 2) and radius 4
d. Center (2, 4) and radius 4
e. Center (4, 2) and radius 0

15. If the ordered pair $(-3, -4)$ is reflected over the *x*-axis, what's the new ordered pair?

a. $(-3, -4)$
b. $(3, -4)$
c. $(3, 4)$
d. $(-3, 4)$
e. $(-4, -3)$

16. What's the midpoint of a line segment with endpoints $(-1, 2)$ and $(3, -6)$?

a. $(1, 2)$
b. $(1, 0)$
c. $(-1, 2)$
d. $(1, -2)$
e. $(1, 4)$

17. Given the following triangle, what's the length of the missing side? Round the answer to the nearest tenth.

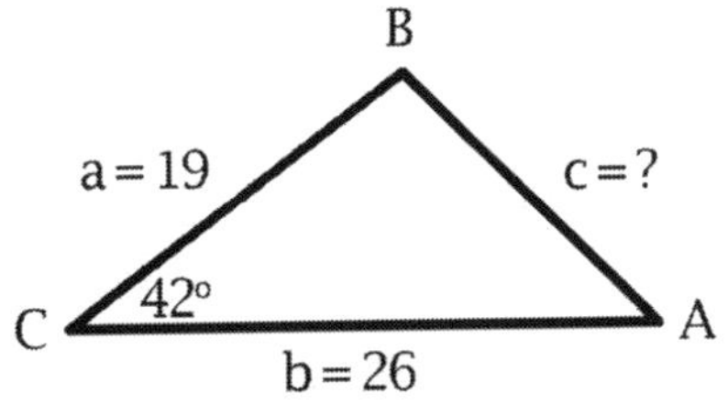


a. 17.0
b. 17.4
c. 18.0
d. 18.4
e. 17.2

18. What are the coordinates of the focus of the parabola $y = -9x^2$?

a. $(-3, 0)$
b. $\left(-\frac{1}{36}, 0\right)$
c. $(0, -3)$
d. $\left(-3, -\frac{1}{36}\right)$
e. $\left(0, -\frac{1}{36}\right)$

19. According to building code regulations, the roof of a house has to be set at a minimum angle of 39° up to a maximum angle of 48° to ensure snow and rain will properly slide off it. What is the maximum incline in terms of radians?

a. $\frac{\pi}{4}$
b. $\frac{\pi}{15}$
c. $\frac{4\pi}{15}$
d. $\frac{3\pi}{4}$
e. $\frac{48\pi}{90}$

20. An arc is intercepted by a central angle of 240°. What is the number of radians of that angle?

a. $\frac{3\pi}{4}$

b. $\frac{4\pi}{3}$

c. $\frac{\pi}{4}$

d. $\frac{\pi}{3}$

e. $\frac{3}{4\pi}$

## Answer Explanations

**1. C:** The triangle in Choice *B* doesn't contain a line of symmetry. The figures in Choices *A* and *D* do contain a line of symmetry but it is not the line that is shown here. Choice *C* is the only one with a correct line of symmetry shown, such that the figure is mirrored on each side of the line.

**2. D:** The slope is given by the change in *y* divided by the change in *x*. The change in *y* is $2 - 0 = 2$, and the change in *x* is $0 - (-4) = 4$. The slope is $\frac{2}{4} = \frac{1}{2}$.

**3. B:** The graph of $y = x^2 + 2$ has a vertex at (0, 2) on the y-axis. The circle with a center at (0, 1) also lies on the y-axis. With a radius of 1, the circle touches the parabola at one point. The vertex of the parabola is (0, 2).

**4. B:** The volume of a rectangular prism is the $length \times width \times height$, and $3cm \times 5cm \times 11cm$ is 165 cm$^3$. Choice *A* is not the correct answer because that is $3cm + 5cm + 11cm$. Choice *C* is not the correct answer because that is 15$^2$. Choice *D* is not the correct answer because that is $3cm \times 5cm \times 10cm$. Lastly, Choice *E* is incorrect because it is $(3cm + 5cm) \times 11cm$.

**5. D:** The volume of a pyramid is $(length \times width \times height)$, divided by 3, and $(12 \times 15)$, divided by 3 is 60 in$^3$. Choice *A* is not the correct answer because that is $12 \times 15$. Choice *B* is not the correct answer because that is $(12 \times 15)$, divided by 2. Choice *C* is not the correct answer because that is $15 \times 2$. Choice *E* is not the correct answer because that is $(12 \times 15)$, divided by 4.

**6. E:** The formula for the area of the circle is $\pi r^2$ and 9 squared is 81. Choice *A* is not the correct answer because that takes the square root of the radius instead of squaring the radius. Choice *B* is not the correct answer because that is $2 \times 9$. Choice *C* is not the correct answer because that is $9 \times 10$. Choice *D* is not the correct answer because that is simply the value of the radius.

**7. B:** The inscribed rectangle is $9 \times 12$ inches. First find the length of *AC* using the Pythagorean Theorem. So, $9^2 + 12^2 = c^2$, where *c* is the length of *AC* in this case. This means that $AC = 15$ inches. This means the diameter of the circle is 15 inches. This can be used to find the area of the entire circle. The formula is $\pi r^2$. So, $3.14(7.5)^2 = 176.6$ sq. inches. Then, subtract the area of the rectangle to find just the area of the shaded region. This is $176.6 - 108 = 68.6$.

**8. C:** Because the triangles are similar, the lengths of the corresponding sides are proportional. Therefore:

$$\frac{30 + x}{30} = \frac{22}{14} = \frac{y + 15}{y}$$

This results in the equation $14(30 + x) = 22 \times 30$ which, when solved, gives $x = 17.1$. The proportion also results in the equation $14(y + 15) = 22y$ which, when solved, gives $y = 26.3$.

**9. A:** Because the volume of the given sphere is $288\pi$ cubic meters, this means:

$$\frac{4}{3}\pi r^3 = 288\pi$$

This equation is solved for $r$ to obtain a radius of 6 meters. The formula for the surface area of a sphere is $4\pi r^2$, so if $r = 6$ in this formula, the surface area is $144\pi$ square meters.

**10. E:** SOHCAHTOA is used to find the missing side length. Because the angle and adjacent side are known, $\tan 60 = \frac{x}{13}$. Making sure to evaluate tangent with an argument in degrees, this equation gives:

$$x = 13 \tan 60 = 13 \times \sqrt{3} = 22.52$$

**11. E:** To calculate the circumference of a circle, use the formula $2\pi r$, where $r$ equals the radius or half of the diameter of the circle and $\pi = 3.14 \ldots$. Substitute the given information, $2\pi 5 = 31.4 \ldots$, answer *E*.

**12. C:** The measure of two complementary angles sums up to 90 degrees. $90 - 54 = 36$. Therefore, the complementary angle is 36 degrees.

**13. B:** To solve this, first find the area of one of the sides. Then multiply times 4 since all sides are identical. To find the area of a side use the formula $\frac{1}{2}bh$, where *b* is the base and *h* is the height of the triangle. For this it is given that the base is 8 cm. To calculate the height, use the Pythagorean Theorem. So, make a smaller triangle within this one where the height will be the unknown side. So, $4^2 + b^2 = 8^2$, $64 - 16 = 48$, $\sqrt{48} = 6.92$. The height of this triangle is 6.92 cm. Now plug this into the formula for area. So:

$$\frac{1}{2}(8)(6.92) = 27.68$$

Then multiply times 4 to get 110.7 as the total surface area.

**14. B:** The technique of completing the square must be used to change $2x^2 + 2y^2 - 16x - 8y + 32 = 0$ into the standard equation of a circle. First, the constant must be moved to the right-hand side of the equals sign, and each term must be divided by the coefficient of the $x^2$ term (which is 2). The $x$ and $y$ terms must be grouped together to obtain:

$$x^2 - 8x + y^2 - 4y = -16$$

Then, the process of completing the square must be completed for each variable. This gives:

$$(x^2 - 8x + 16) + (y^2 - 4y + 4) = -16 + 16 + 4$$

The equation can be written as:

$$(x - 4)^2 + (y - 2)^2 = 4$$

Therefore, the center of the circle is (4, 2) and the radius is:

$$\sqrt{4} = 2$$

**15. D:** When an ordered pair is reflected over an axis, the sign of one of the coordinates must change. When it's reflected over the $x$-axis, the sign of the $x$ coordinate must change. The *y*-value remains the same. Therefore, the new ordered pair is $(-3, 4)$.

**16. D:** The midpoint formula should be used.

$$M = \left(\frac{x_1 + x_2}{2}, \frac{y_1 + y_2}{2}\right) = \left(\frac{-1 + 3}{2}, \frac{2 + (-6)}{2}\right) = (1, -2)$$

**17. B:** Because this isn't a right triangle, SOHCAHTOA can't be used. However, the law of cosines can be used. Therefore:

$$c^2 = a^2 + b^2 - 2ab$$

$$\cos C = 19^2 + 26^2 - 2 \cdot 19 \cdot 26 \cdot \cos 42°$$

$$302.773$$

Taking the square root and rounding to the nearest tenth results in $c = 17.4$.

**18. E:** A parabola of the form $y = \frac{1}{4f}x^2$ has a focus $(0, f)$. Because $y = -9x^2$, set $-9 = \frac{1}{4f}$. Solving this equation for $f$ results in $f = -\frac{1}{36}$. Therefore, the coordinates of the focus are $\left(0, -\frac{1}{36}\right)$.

**19. C:** When you simplify $\frac{48° \times \pi}{180}$, you get $\frac{4\pi}{15}$. Choice *A* is not the correct answer because $\frac{\pi}{4}$ is 45°. Choice *B* is not the correct answer because $\frac{\pi}{15}$ is 12°. Choice *D* is not the correct answer because $\frac{3\pi}{4} = 135°$. Choice *E* is not the correct answer because the initial equation was set the incorrect denominator (90 instead of 180) and failed to simplify.

**20. B:** When $(240) \times (\frac{\pi}{180})$ is simplified, the result is $\frac{4\pi}{3}$. Choice *A* is not the correct answer because the numbers in the numerator and the denominator are switched. Choice *C* is not the correct answer because it is not the correct reduction of the fraction. Choice *D* is not the correct answer because it is not the correct reduction of the fraction. Choice *E* is incorrect because it is the reciprocal of the correct answer.

# Data Analysis, Statistics, and Probability

## Statistical Questions

A statistical question is answered by collecting data with variability. Data consists of facts and/or statistics (numbers), and variability refers to a tendency to shift or change. Data is a broad term, inclusive of things like height, favorite color, name, salary, temperature, gas mileage, and language. Questions requiring data as an answer are not necessarily statistical questions. If there is no variability in the data, then the question is not statistical in nature. Consider the following examples: what is Mary's favorite color? How much money does your mother make? What was the highest temperature last week? How many miles did your car get on its last tank of gas? How much taller than Bob is Ed?

None of the above are statistical questions because each case lacks variability in the data needed to answer the question. The questions on favorite color, salary, and gas mileage each require a single piece of data, whether a fact or statistic. Therefore, variability is absent. Although the temperature question requires multiple pieces of data (the high temperature for each day), a single, distinct number is the answer. The height question requires two pieces of data, Bob's height and Ed's height, but no difference in variability exists between those two values. Therefore, this is not a statistical question. Statistical questions typically require calculations with data.

Consider the following statistical questions:

*How many miles per gallon of gas does the 2016 Honda Civic get?* To answer this question, data must be collected. This data should include miles driven and gallons used. Different cars, different drivers, and different driving conditions will produce different results. Therefore, variability exists in the data. To answer the question, the mean (average) value could be determined.

*Are American men taller than German men?* To answer this question, data must be collected. This data should include the heights of American men and the heights of German men. All American men are not the same height and all German men are not the same height. Some American men are taller than some German men and some German men are taller than some American men. Therefore, variability exists in the data. To answer the question, the median values for each group could be determined and compared.

The following are more examples of statistical questions: What proportion of $4^{th}$ graders have a favorite color of blue? How much money do teachers make? Is it colder in Boston or Chicago?

## Mean, Median, and Mode

### Mean

Suppose that you have a set of data points and some description of the general properties of this data need to be found.

The first property that can be defined for this set of data is the *mean*. This is the same as average. To find the mean, add up all the data points, then divide by the total number of data points. For example, suppose that in a class of 10 students, the scores on a test were 50, 60, 65, 65, 75, 80, 85, 85, 90, 100. Therefore, the average test score will be:

$$\frac{50 + 60 + 65 + 65 + 75 + 80 + 85 + 85 + 90 + 100}{10} = 75.5$$

The mean is a useful number if the distribution of data is normal (more on this later), which roughly means that the frequency of different outcomes has a single peak and is roughly equally distributed on both sides of that peak. However, it is less useful in some cases where the data might be split or where there are some *outliers*. Outliers are data points that are far from the rest of the data. For example, suppose there are 10 executives and 90 employees at a company. The executives make $1000 per hour, and the employees make $10 per hour.

Therefore, the average pay rate will be:

$$\frac{\$1000 \times 10 + \$10 \times 90}{100} = \$109 \text{ per hour}$$

In this case, this average is not very descriptive since it's not close to the actual pay of the executives or the employees.

## Median

Another useful measurement is the *median*. In a data set, the median is the point in the middle. The middle refers to the point where half the data comes before it and half comes after, when the data is recorded in numerical order. For instance, these are the speeds of the fastball of a pitcher during the last inning that he pitched (in order from least to greatest):

90, 92, 93, 93, 95, 96, 97, 97, 97

There are nine total numbers, so the middle or *median* number is the 5th one, which is 95.

In cases where the number of data points is an even number, then the average of the two middle points is taken. In the previous example of test scores, the two middle points are 75 and 80. Since there is no single point, the average of these two scores needs to be found. The average is:

$$\frac{75 + 80}{2} = 77.5$$

The median is generally a good value to use if there are a few outliers in the data. It prevents those outliers from affecting the "middle" value as much as when using the mean.

Since an outlier is a data point that is far from most of the other data points in a data set, this means an outlier also is any point that is far from the median of the data set. The outliers can have a substantial effect on the mean of a data set, but they usually do not change the median or mode, or do not change them by a large quantity. For example, consider the data set (3, 5, 6, 6, 6, 8). This has a median of 6 and a mode of 6, with a mean of $\frac{34}{6} \approx 5.67$. Now, suppose a new data point of 1000 is added so that the data set is now (3, 5, 6, 6, 6, 8, 1000). This does not change the median or mode, which are both still 6. However, the average is now $\frac{1034}{7}$, which is approximately 147.7. In this case, the median and mode will be better descriptions for most of the data points.

The reason for outliers in a given data set is a complicated problem. It is sometimes the result of an error by the experimenter, but often they are perfectly valid data points that must be taken into consideration.

## Mode

One additional measure to define for $X$ is the *mode*. This is the data point that appears most frequently. If two or more data points all tie for the most frequent appearance, then each of them is considered a mode. In the case of the test scores, where the numbers were 50, 60, 65, 65, 75, 80, 85, 85, 90, 100, there are two modes: 65 and 85.

## Normal Distribution

A *normal distribution* of data follows the shape of a bell curve. In a normal distribution, the data set's median, mean, and mode are equal. Therefore, 50 percent of its values are less than the mean and 50 percent are greater than the mean. Data sets that follow this shape can be generalized using normal distributions. Normal distributions are described as *frequency distributions* in which the data set is plotted as percentages rather than true data points. A *relative frequency distribution* is one where the y-axis is between zero and 1, which is the same as 0% to 100%. Within a standard deviation, 68 percent of the values are within 1 standard deviation of the mean, 95 percent of the values are within 2 standard deviations of the mean, and 99.7 percent of the values are within 3 standard deviations of the mean. The number of standard deviations that a data point falls from the mean is called the *z-score*. The formula for the z-score is $Z = \frac{x-\mu}{\sigma}$, where $\mu$ is the mean, $\sigma$ is the standard deviation, and $x$ is the data point. This formula is used to fit any data set that resembles a normal distribution to a standard normal distribution, in a process known as *standardizing*. Here is a normal distribution with labelled z-scores:

**Normal Distribution with Labelled Z-Scores**

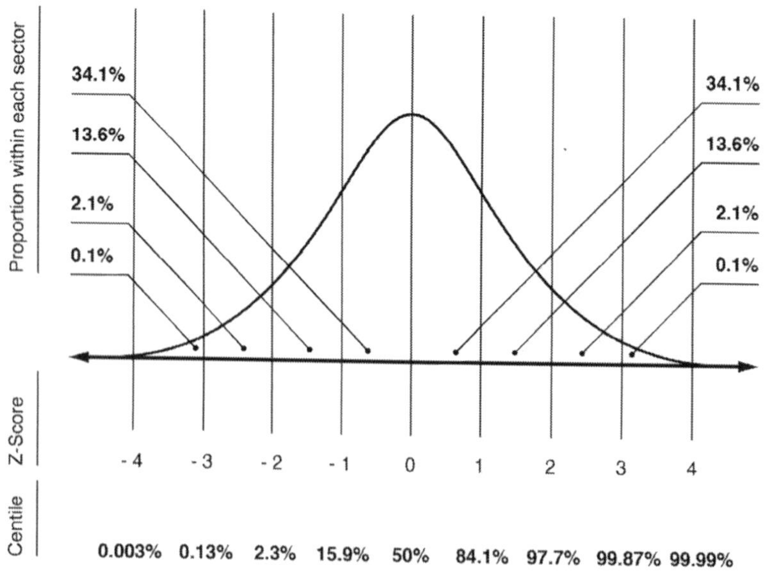

Population percentages can be estimated using normal distributions. For example, the probability that a data point will be less than the mean, or that the z-score will be less than 0, is 50%. Similarly, the probability that a data point will be within 1 standard deviation of the mean, or that the z-score will be between -1 and 1, is about 68.2%. When using a table, the left column states how many standard deviations (to one decimal place) away from the mean the point is, and the row heading states the second decimal place. The entries in the table corresponding to each column and row give the probability, which is equal to the area.

## Best Description of a Set of Data

Measures of central tendency, namely mean, median, and mode, describe characteristics of a set of data. Specifically, they are intended to represent a *typical* value in the set by identifying a central position of the set. Depending on the characteristics of a specific set of data, different measures of central tendency are more indicative of a typical value in the set.

When a data set is grouped closely together with a relatively small range and the data is spread out somewhat evenly, the mean is an effective indicator of a typical value in the set. Consider the following data set representing the height of sixth grade boys in inches: 61 inches, 54 inches, 58 inches, 63 inches, 58 inches. The mean of the set is 58.8 inches. The data set is grouped closely (the range is only 9 inches) and the values are spread relatively evenly (three values below the mean and two values above the mean). Therefore, the mean value of 58.8 inches is an effective measure of central tendency in this case.

When a data set contains a small number of values either extremely large or extremely small when compared to the other values, the mean is not an effective measure of central tendency. Consider the following data set representing annual incomes of homeowners on a given street: $71,000; $74,000; $75,000; $77,000; $340,000. The mean of this set is $127,400. This figure does not indicate a typical value in the set, which contains four out of five values between $71,000 and $77,000. The median is a much more effective measure of central tendency for data sets such as these. Finding the middle value diminishes the influence of outliers, or numbers that may appear out of place, like the $340,000 annual income. The median for this set is $75,000 which is much more typical of a value in the set.

The mode of a data set is a useful measure of central tendency for categorical data when each piece of data is an option from a category. Consider a survey of 31 commuters asking how they get to work with results summarized below.

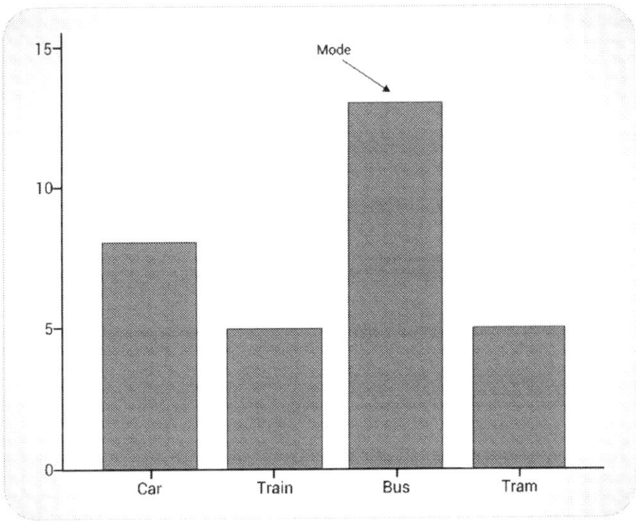

The mode for this set represents the value, or option, of the data that repeats most often. This indicates that the bus is the most popular method of transportation for the commuters.

## Effects of Changes in Data

Changing all values of a data set in a consistent way produces predictable changes in the measures of the center and range of the set. A linear transformation changes the original value into the new value by either adding a given number to each value, multiplying each value by a given number, or both. Adding (or subtracting) a given value to each data point will increase (or decrease) the mean, median, and any modes by the same value. However, the range will remain the same due to the way that range is calculated. Multiplying (or dividing) a given value by each data point will increase (or decrease) the mean, median, and any modes, and the range by the same factor.

Consider the following data set, call it set *P*, representing the price of different cases of soda at a grocery store: $4.25, $4.40, $4.75, $4.95, $4.95, $5.15. The mean of set *P* is $4.74. The median is $4.85. The mode of the set is $4.95. The range is $0.90. Suppose the state passes a new tax of $0.25 on every case of soda sold. The new data set, set *T*, is calculated by adding $0.25 to each data point from set *P*. Therefore, set *T* consists of the following values: $4.50, $4.65, $5.00, $5.20, $5.20, $5.40. The mean of set *T* is $4.99. The median is $5.10. The mode of the set is $5.20. The range is $.90. The mean, median and mode of set *T* is equal to $0.25 added to the mean, median, and mode of set *P*. The range stays the same.

Now suppose, due to inflation, the store raises the cost of every item by 10 percent. Raising costs by 10 percent is calculated by multiplying each value by 1.1. The new data set, set *I*, is calculated by multiplying each data point from set *T* by 1.1. Therefore, set *I* consists of the following values: $4.95, $5.12, $5.50, $5.72, $5.72, $5.94. The mean of set *I* is $5.49. The median is $5.61. The mode of the set is $5.72. The range is $0.99. The mean, median, mode, and range of set *I* is equal to 1.1 multiplied by the mean, median, mode, and range of set *T* because each increased by a factor of 10 percent.

## Comparing Data

Data sets can be compared by looking at the center and spread of each set. Measures of central tendency involve median, mean, midrange, and mode. The *mode* of a data set is the data value or values that appears the most frequently. The *midrange* is equal to the maximum value plus the minimum value divided by two. The *median* is the value that is halfway into each data set; it splits the data into two intervals. The *mean* is the sum of all data values divided by the number of data points. Two completely different sets of data can have the same mean. For example, a data set having values ranging from 0 to 100 and a data set having values ranging from 44 to 46 could both have means equal to 50. The first data set would have a much wider range, which is known as the *spread* of the data. It measures how varied the data is within each set. Spread can be defined further as either interquartile range or standard deviation. The *interquartile range (IQR)* is the range of the middle fifty percent of the data set. The *standard deviation, s,* quantifies the amount of variation with respect to the mean. A lower standard deviation shows that the data set does not differ much from the mean. A larger standard deviation shows that the data set is spread out farther away from the mean. The formula for standard deviation is:

$$s = \sqrt{\frac{\sum(x - \bar{x})^2}{n - 1}}$$

where $x$ is each value in the data set, $\bar{x}$ is the mean, and $n$ is the total number of data points in the sample set. The square of the standard deviation is known as the *variance* of the data set. A data set can have outliers, and measures of central tendency that are not affected by outliers are the mode and median. Those measures are labeled as resistant measures of center.

## Range, Interquartile Range, and Spread of Data

A set of data can be described in terms of its center, spread, shape and any unusual features. The center of a data set can be measured by its mean, median, or mode. The spread, or variability, of a data set refers to how far the data points are from the center (mean or median). The spread can be measured by the range or the quartiles and interquartile range. A data set with data points clustered around the center will have a small spread. A data set covering a wide range will have a large spread. Essentially, the spread of a data set refers to how far the data points are from the center (mean or median). The spread can be measured by the range or the quartiles and interquartile range. A data set with data points clustered around the center will have a small spread. A data set covering a wide range will have a large spread. The **interquartile range** *(IQR)* is the range of the middle 50 percent of the data set. This range can be seen in the large rectangle on a box plot.

When a data set is displayed as a **histogram** or frequency distribution plot, the shape indicates if a sample is normally distributed, symmetrical, or has measures of skewness or kurtosis.

When graphed, a data set with a **normal distribution** will resemble a bell curve.

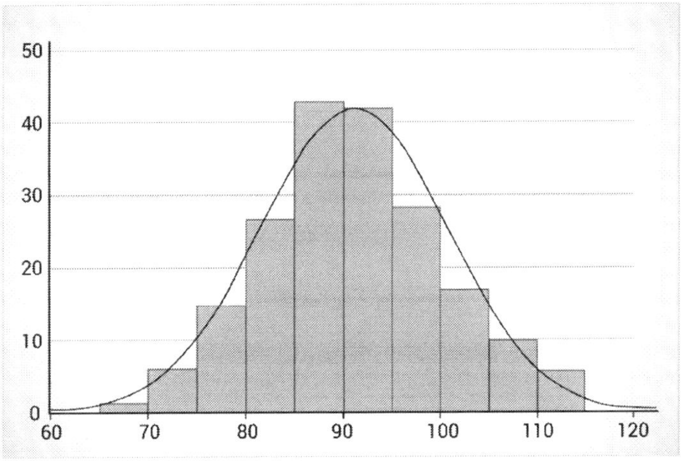


If the data set is symmetrical, each half of the graph when divided at the center is a mirror image of the other. If the graph has fewer data points to the right, the data is **skewed right**. If it has fewer data points to the left, the data is **skewed left**.

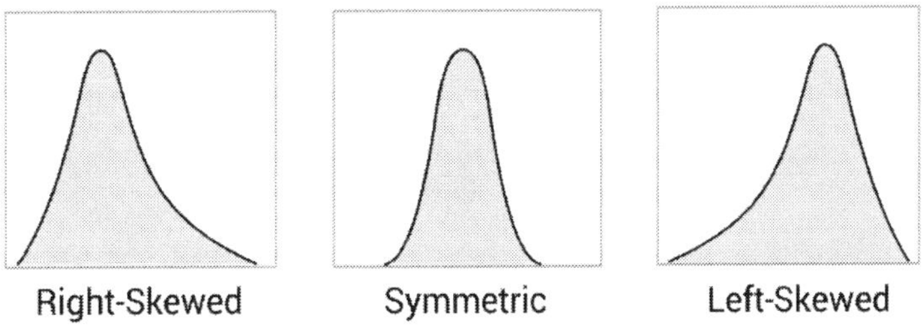

**Kurtosis** is a measure of whether the data is heavy-tailed with a high number of outliers, or light-tailed with a low number of outliers.

A description of a data set should include any unusual features such as gaps or outliers. A **gap** is a span within the range of the data set containing no data points. An **outlier** is a data point with a value either extremely large or extremely small when compared to the other values in the set.

The **range** of a data set is the difference between the highest and the lowest values in the set. The range can be considered to be the span of the data set. To determine the range, the smallest value in the set is subtracted from the largest value. The ranges for the data sets A, B, and C above are calculated as follows:

A: $14 - 7 = 7$

B: $51 - 33 = 18$

C: $173 - 151 = 22$

## Quartiles and Percentiles

The *first quartile* of a set of data *X* refers to the largest value from the first ¼ of the data points. In practice, there are sometimes slightly different definitions that can be used, such as the median of the first half of the data points (excluding the median itself if there are an odd number of data points). The term also has a slightly different use: when it is said that a data point lies *in the first quartile*, it means it is less than or equal to the median of the first half of the data points. Conversely, if it lies *at* the first quartile, then it is equal to the first quartile.

When it is said that a data point lies in the *second quartile*, it means it is between the first quartile and the median.

The *third quartile* refers to data that lies between ½ and ¾ of the way through the data set. Again, there are various methods for defining this precisely, but the simplest way is to include all of the data that lie between the median and the median of the top half of the data.

Data that lies in the *fourth quartile* refers to all of the data above the third quartile.

*Percentiles* may be defined in a similar manner to quartiles. Generally, this is defined in the following manner:

If a data point lies *in the n-th percentile*, this means it lies in the range of the first *n*% of the data.

If a data point lies *at* the *n*-th percentile, then it means that *n*% of the data lies below this data point.

## Standard Deviation

Given a data set *X* consisting of data points $(x_1, x_2, x_3, \ldots x_n)$, the *variance* of *X* is defined to be:

$$\frac{\sum_{i=1}^{n}(x_i - \bar{X})^2}{n}$$

This means that the variance of *X* is the average of the squares of the differences between each data point and the mean of *X*.

Given a data set $X$ consisting of data points $(x_1, x_2, x_3, \ldots x_n)$, the *standard deviation* of $X$ is defined to be:

$$s_x = \sqrt{\frac{\sum_{i=1}^{n}(x_i - \bar{X})^2}{n}}$$

In other words, the standard deviation is the square root of the variance.

Both the variance and the standard deviation are measures of how much the data tend to be spread out. When the standard deviation is low, the data points are mostly clustered around the mean. When the standard deviation is high, this generally indicates that the data are quite spread out, or else that there are a few substantial outliers.

As a simple example, compute the standard deviation for the data set (1, 3, 3, 5). First, compute the mean, which will be:

$$\frac{1+3+3+5}{4} = \frac{12}{4} = 3$$

Now, find the variance of $X$ with the formula:

$$\sum_{i=1}^{4}(x_i - \bar{X})^2 = (1-3)^2 + (3-3)^2 + (3-3)^2 + (5-3)^2$$

$$-2^2 + 0^2 + 0^2 + 2^2 = 8$$

Therefore, the variance is $\frac{8}{4} = 2$. Taking the square root, the standard deviation will be $\sqrt{2}$.

Note that the standard deviation only depends upon the mean, not upon the median or mode(s). Generally, if there are multiple modes that are far apart from one another, the standard deviation will be high. A high standard deviation does not always mean there are multiple modes, however.

## Graphs and Plots

Various graphs can be used to visually represent a given set of data. Each type of graph requires a different method of arranging data points and different calculations of the data. To construct a histogram, the range of the data points is divided into equal intervals. The frequency for each interval is then determined, which reveals how many points fall into each interval. A graph is constructed with the vertical axis representing the frequency and the horizontal axis representing the intervals. The lower value of each interval should be labeled along the horizontal axis. Finally, for each interval, a bar is drawn from the lower value of each interval to the lower value of the next interval with a height equal to the frequency of the interval. Because of the intervals, histograms do not have any gaps between bars along the horizontal axis.

To construct a box (or box-and-whisker) plot, the five number summary for the data set is calculated as follows: the **second quartile ($Q_2$)** is the median of the set. The **first quartile ($Q_1$)** is the median of the values below $Q_2$. The **third quartile ($Q_3$)** is the median of the values above $Q_2$. The **upper extreme** is the highest value in the data set if it is not an **outlier** (greater than 1.5 times the interquartile range $Q_3$ - $Q_1$). The lower extreme is the least value in the data set if it is not an outlier (more than 1.5 times lower than the interquartile range). To construct the box-and-whisker plot, each value is plotted on a number line, along with any outliers. The **box** consists of $Q_1$ and $Q_3$ as its *top* and *bottom* and $Q_2$ as the dividing line inside the box. The **whiskers** extend from the lower extreme to $Q_1$ and from $Q_3$ to the upper extreme.

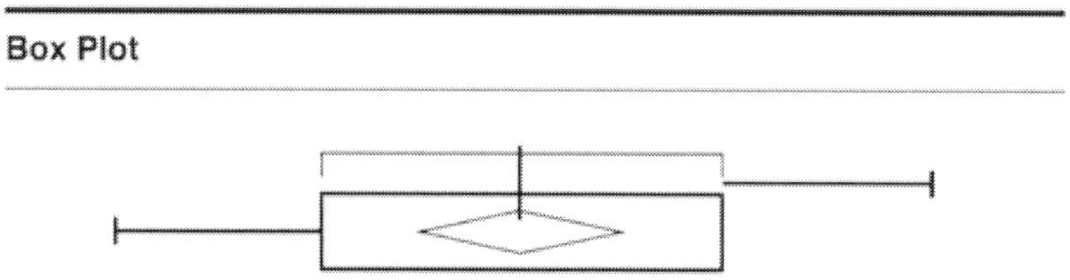


A scatter plot displays the relationship between two variables. Values for the independent variable, typically denoted by *x*, are paired with values for the dependent variable, typically denoted by *y*. Each set of corresponding values are written as an ordered pair (*x*, *y*). To construct the graph, a coordinate grid is labeled with the *x*-axis representing the independent variable and the *y*-axis representing the dependent variable. Each ordered pair is graphed.

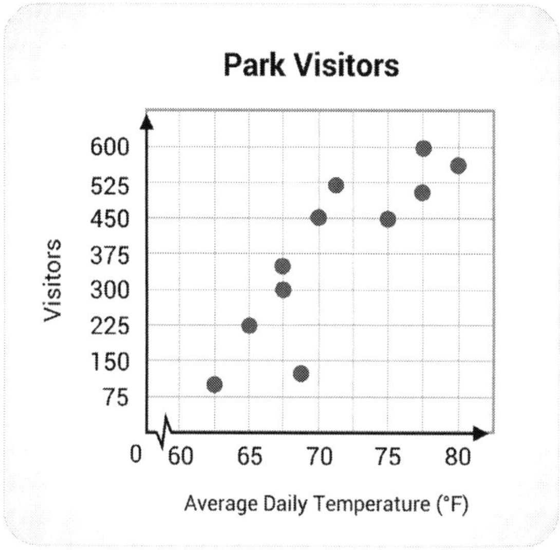

Like a scatter plot, a **line graph** compares variables that change continuously, typically over time. Paired data values (ordered pair) are plotted on a coordinate grid with the *x*- and *y*-axis representing the variables. A line is drawn from each point to the next, going from left to right. The line graph below displays cell phone use for given years (two variables) for men, women, and both sexes (three data sets).

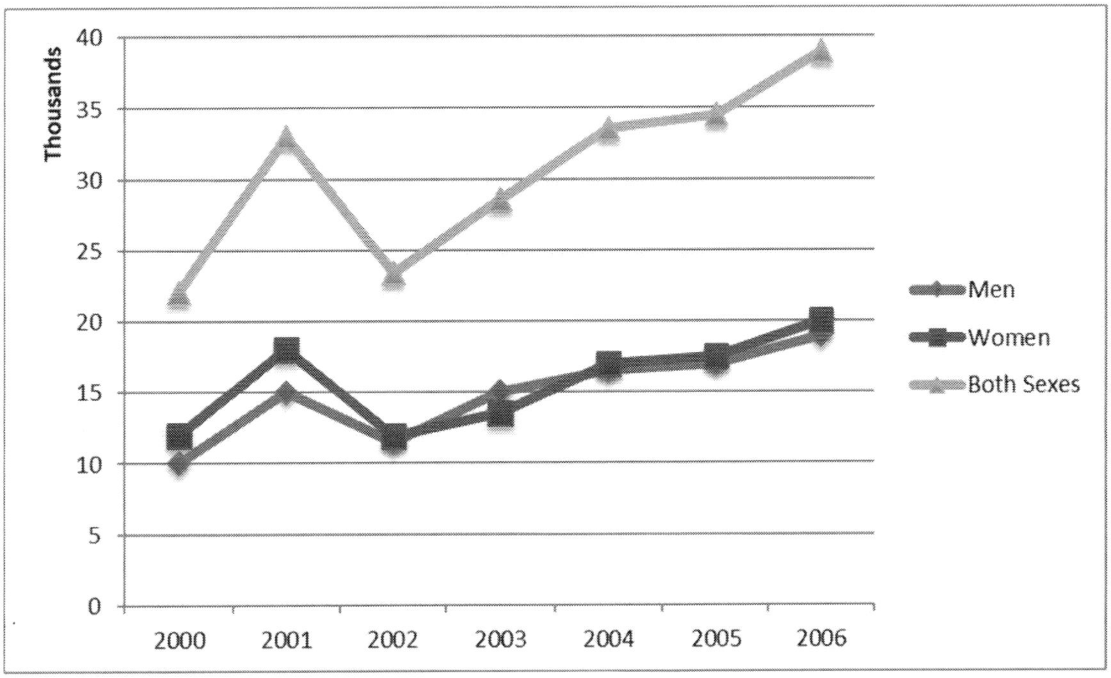


A **line plot**, also called **dot plot**, displays the frequency of data (numerical values) on a number line. To construct a line plot, a number line is used that includes all unique data values. It is marked with x's or dots above the value the number of times that the value occurs in the data set.

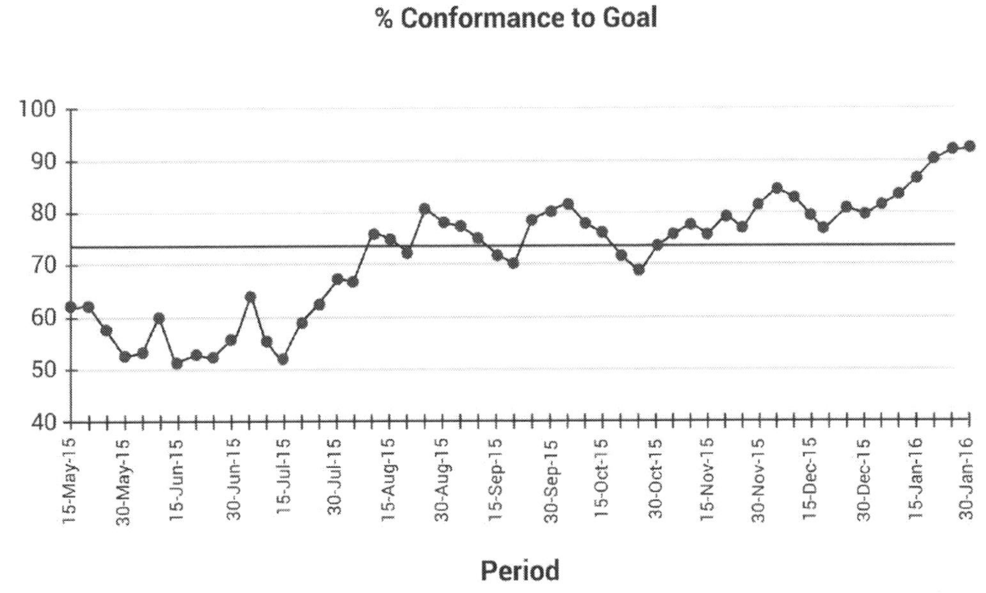

A **picture graph** is a diagram that shows pictorial representation of data being discussed. The symbols utilized can represent the quantity of the objects being discussed. By using the key, it should be determined that each fruit symbol in the following graph represents a count of two fruits. One drawback of picture graphs is that they can be less accurate if each symbol represents a large number.

For example, if each banana symbol represented ten bananas, and students consumed 22 bananas, it may be challenging to draw and interpret two and one-fifth bananas as a frequency count of 22.

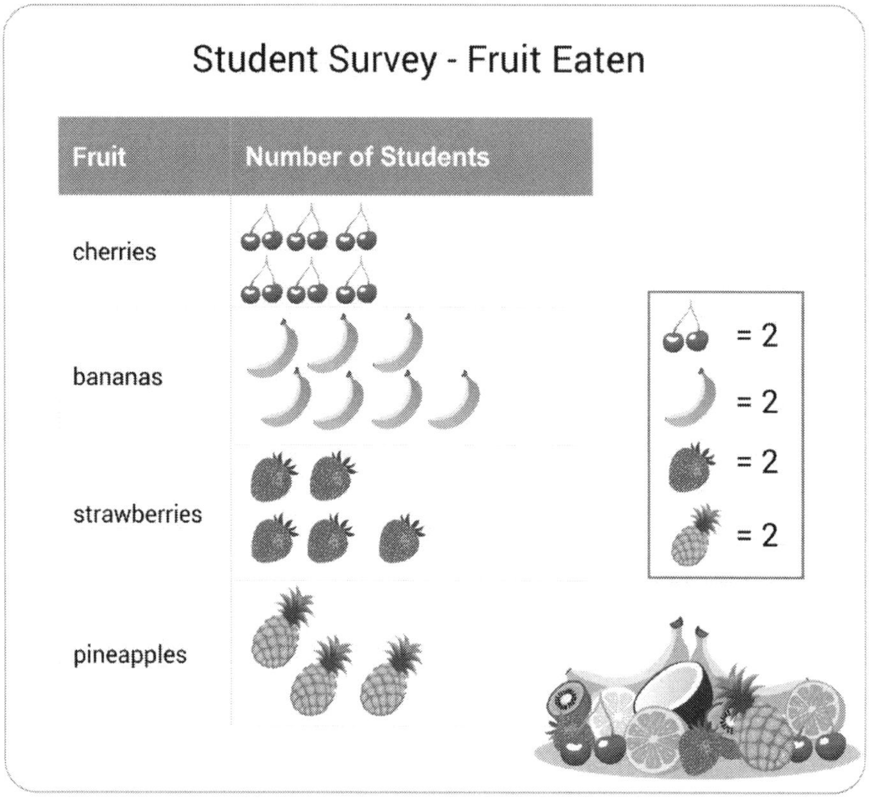


A **stem-and-leaf plot** is a method of displaying sets of data by organizing the numbers by their stems (usually the tens digit) and the different leaf values (usually the ones digit).

For example, to organize a number of movie critic's ratings, as listed below, a stem and leaf plot could be utilized to display the information in a more condensed manner.

Movie critic scores: 47, 52, 56, 59, 61, 64, 66, 68, 68, 70, 73, 75, 79, 81, 83, 85, 86, 88, 88, 89, 90, 90, 91, 93, 94, 96, 96, 99.

| | Movie Ratings | | | | | | | |
|---|---|---|---|---|---|---|---|---|
| **4** | 7 | | | | | | | |
| **5** | 2 | 6 | 9 | | | | | |
| **6** | 1 | 4 | 6 | 8 | 8 | | | |
| **7** | 0 | 3 | 5 | 9 | | | | |
| **8** | 1 | 3 | 5 | 6 | 8 | 8 | 9 | |
| **9** | 0 | 0 | 1 | 3 | 4 | 6 | 6 | 9 |
| **Key** | 6 \| 1 represents 61 | | | | | | | |

Looking at this stem and leaf plot, it is easy to ascertain key features of the data set. For example, what is the range of the data in the stem-and-leaf plot?

Using this method, it is easier to visualize the distribution of the scores and answer the question pertaining to the range of scores, which is $99 - 47 = 52$.

A **tally chart** is a diagram in which tally marks are utilized to represent data. Tally marks are a means of showing a quantity of objects within a specific classification. Here is an example of a tally chart:

| Number of days with rain | Number of weeks |
|---|---|
| 0 | \|\| |
| 1 | 𝍸 |
| 2 | 𝍸 \| |
| 3 | 𝍸 \|\|\|\| |
| 4 | 𝍸 𝍸 𝍸 |
| 5 | 𝍸 |
| 6 | 𝍸 \| |
| 7 | \|\|\|\| |

Data is often recorded using fractions, such as half a mile, and understanding fractions is critical because of their popular use in real-world applications. Also, it is extremely important to label values with their units when using data. For example, regarding length, the number 2 is meaningless unless it is attached to a unit. Writing 2 cm shows that the number refers to the length of an object.

Another way to represent data is with a **frequency table**. Like a tally chart, these can display the number of times specific answers are given, in order to provide a clearer overall picture of information. The following is the data representing the scores gathered on a patient satisfaction survey, visualized on a frequency dot table.

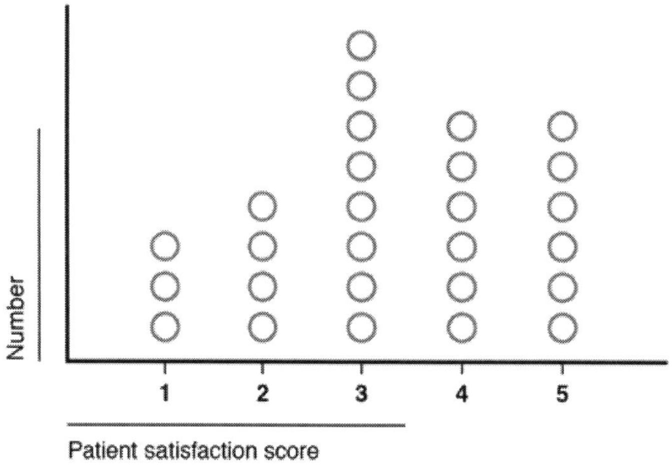


It is clear that the majority of the scores are in the middle, and higher at 2.0 and 2.5.

A **bar graph** is a diagram in which the quantity of items within a specific classification is represented by the height of a rectangle. Each type of classification is represented by a rectangle of equal width. Here is an example of a bar graph:

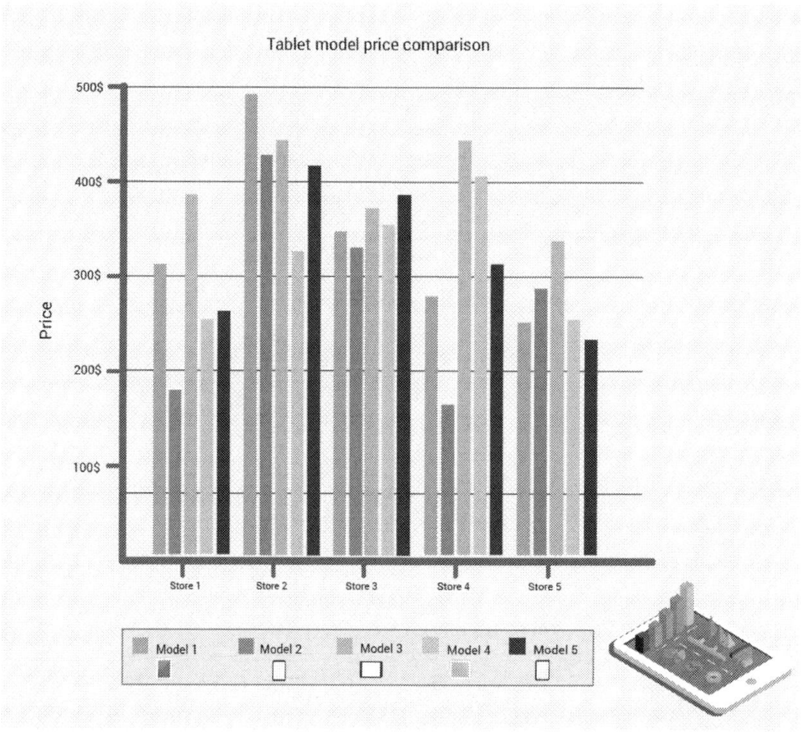


A **circle graph**, also called a **pie chart**, shows categorical data with each category representing a percentage of the whole data set. To make a circle graph, the percent of the data set for each category must be determined. To do so, the frequency of the category is divided by the total number of data points and converted to a percent. For example, if 80 people were asked what their favorite sport is and 20 responded basketball, basketball makes up 25% of the data ($\frac{20}{80}$ =.25=25%). Each category in a data set is represented by a slice of the circle proportionate to its percentage of the whole.

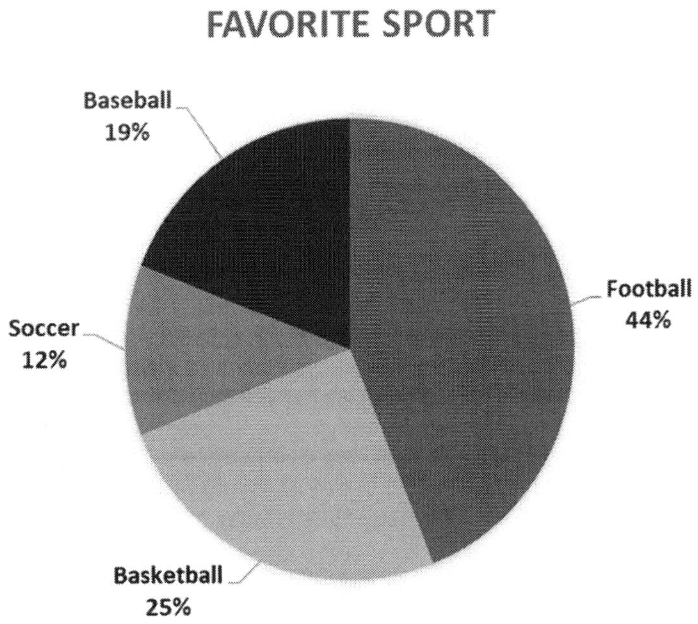


## Choosing an Appropriate Graph Based on Data

Choosing the appropriate graph to display a data set depends on what type of data is included in the set and what information must be displayed. Histograms and box plots can be used for data sets consisting of individual values across a wide range. Examples include test scores and incomes. Histograms and box plots will indicate the center, spread, range, and outliers of a data set. A histogram will show the shape of the data set, while a box plot will divide the set into quartiles (25% increments), allowing for comparison between a given value and the entire set.

Scatter plots and line graphs can be used to display data consisting of two variables. Examples include height and weight, or distance and time. A correlation between the variables is determined by examining the points on the graph. Line graphs are used if each value for one variable pairs with a distinct value for the other variable. Line graphs show relationships between variables.

Line plots, bar graphs, and circle graphs are all used to display categorical data, such as surveys. Line plots and bar graphs both indicate the frequency of each category within the data set. A line plot is used when the categories consist of numerical values. For example, the number of hours of TV watched by individuals is displayed on a line plot. A bar graph is used when the categories consists of words. For example, the favorite ice cream of individuals is displayed with a bar graph. A circle graph can be used to display either type of categorical data. However, unlike line plots and bar graphs, a circle graph does not indicate the frequency of each category. Instead, the circle graph represents each category as its percentage of the whole data set.

## Drawing Conclusions Based on Graphical Displays

In an experiment, variables are the key to analyzing data, especially when data is in a graph or table. Variables can represent anything, including objects, conditions, events, and amounts of time.

**Covariance** is a general term referring to how two variables move in relation to each other. Take for example an employee that gets paid by the hour. For them, hours worked and total pay have a positive covariance. As hours worked increases, so does pay.

**Constant variables** remain unchanged by the scientist across all trials. Because they are held constant for all groups in an experiment, they aren't being measured in the experiment, and they are usually ignored. Constants can either be controlled by the scientist directly like the nutrition, water, and sunlight given to plants, or they can be selected by the scientist specifically for an experiment like using a certain animal species or choosing to investigate only people of a certain age group.

**Independent variables** are also controlled by the scientist, but they are the same only for each group or trial in the experiment. Each group might be composed of students that all have the same color of car or each trial may be run on different soda brands. The independent variable of an experiment is what is being indirectly tested because it effects change in the dependent variables.

**Dependent variables** experience change caused by the independent variable and are what is being measured or observed. For example, college acceptance rates could be a dependent variable of an experiment that sorted a large sample of high school students by an independent variable such as test scores. In this experiment, the scientist groups the high school students by the independent variable (test scores) to see how it affects the dependent variable (their college acceptance rates).

Note that most variables can be held constant in one experiment but independent or dependent in another. For example, when testing how well a fertilizer aids plant growth, its amount of sunlight should be held constant for each group of plants, but if the experiment is being done to determine the proper amount of sunlight a plant should have, the amount of sunlight is an independent variable because it is necessarily changed for each group of plants.

An **X-Y diagram**, also known as a **scatter diagram**, visually displays the relationship between two variables. The independent variable is placed on the *x-axis*, or horizontal axis, and the dependent variable is placed on the *y-axis*, or vertical axis.

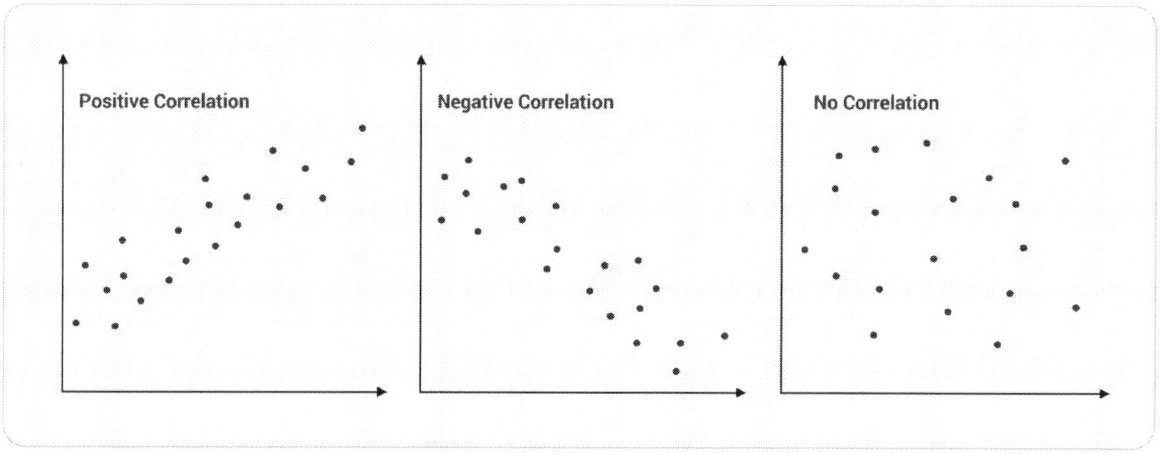

As shown in the figures above, an X-Y diagram may result in positive, negative, or no correlation between the two variables. In the first scatter plot, as the Y factor increases, the X factor increases as well. The opposite is true as well: as the X factor increases the Y factor also increases. Thus there is a positive correlation because one factor appears to positively affect the other factor

It's important to note, however, that a positive correlation between two variables doesn't equate to a cause-and-effect relationship. For example, a positive correlation between labor hours and units produced may not equate to a cause and effect relationship between the two. Any instance of correlation only indicates how likely the presence of one variable is in the instance of another. The variables should be further analyzed to determine which, if any, other variables (i.e. quality of employee work) may contribute to the positive correlation.

Comparing data sets within statistics can mean many things. The first way to compare data sets is by looking at the center and spread of each set. The **center of a data** set can mean two things: median or mean. The **median** is the value that's halfway into each data set, and it splits the data into two intervals. The **mean** is the average value of the data within a set. It's calculated by adding up all of the data in the set and dividing the total by the number of data points. Outliers can significantly impact the mean. Additionally, two completely different data sets can have the same mean. For example, a data set with values ranging from 0 to 100 and a data set with values ranging from 44 to 56 can both have means of 50. The first data set has a much wider range, which is known as the **spread** of the data. This measures how varied the data is within each set.

### *Interpreting Competing Data*

Be careful of questions with competing studies. These questions will ask the student to interpret which of two studies shows the greater amount or the higher rate of change between two results.

Here's an example. A research facility runs studies on two different antibiotics: Drug A and Drug B. The Drug A study includes 1,000 participants and cures 600 people. The Drug B study includes 200 participants and cures 150 people. Which drug is more successful?

The first step is to determine the percentage of each drug's rate of success. Drug A was successful in curing 60% of participants, while Drug B achieved a 75% success rate. Thus, Drug B is more successful based on these studies, even though it cured fewer people.

Sample size and experiment consistency should also be considered when answering questions based on competing studies. Is one study significantly larger than the other? In the antibiotics example, the Drug A study is five times larger than Drug B. Thus, Drug B's higher efficacy (desired result) could be a result of the smaller sample size, rather than the quality of drug.

Consistency between studies is directly related to sample size. Let's say the research facility elects to conduct more studies on Drug B. In the next study, there are 400 participants, and 200 are cured. The success rate of the second study is 50%. The results are clearly inconsistent with the first study, which means more testing is needed to determine the drug's efficacy. A hallmark of mathematical or scientific research is repeatability. Studies should be consistent and repeatable, with an appropriately large sample size, before drawing extensive conclusions.

## Least Squares Regression Models

## Regression Models

*Regression lines* are straight lines that calculate a relationship between nonlinear data involving an independent variable and a dependent variable. A regression line is of the form $y = mx + b$, where $m$ is the slope and $b$ is the $y$-intercept. Both the slope and $y$-intercept are found using the *Method of Least Squares*, which involves minimizing residuals – the difference between the dependent variable from the data set and the estimated value located on the regression line. The slope represents the rate of change in $y$ as $x$ increases. The $y$-intercept is the predicted value when the independent variable is equal to 0. Technology, such as a graphing calculator or Microsoft Excel®, can also be utilized to find the equation of this line. In either case, the data points are entered, and a line is "fit" that best represents the shape of the data. Here is an example of a data set and its regression line:

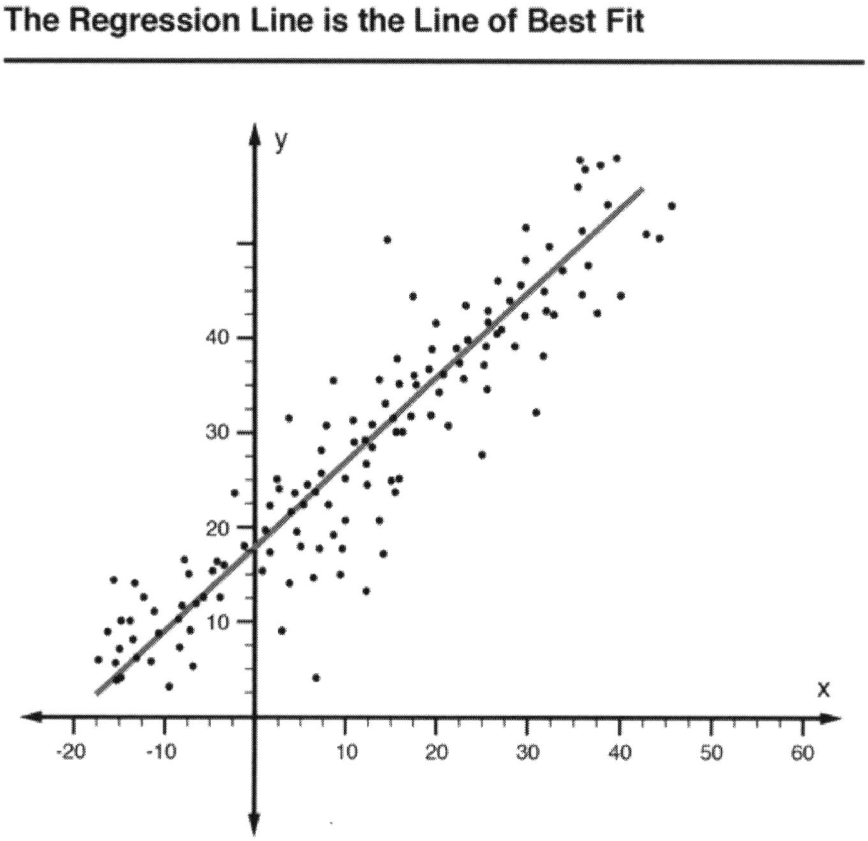


Regression models are highly used for forecasting, and linear regression techniques are the simplest models. If the nonlinear data follows the shape of exponential, logarithmic, or power functions, those types of functions can be used to more accurately model the data rather than lines.

Here is an example of both an exponential regression and a logarithmic regression model:

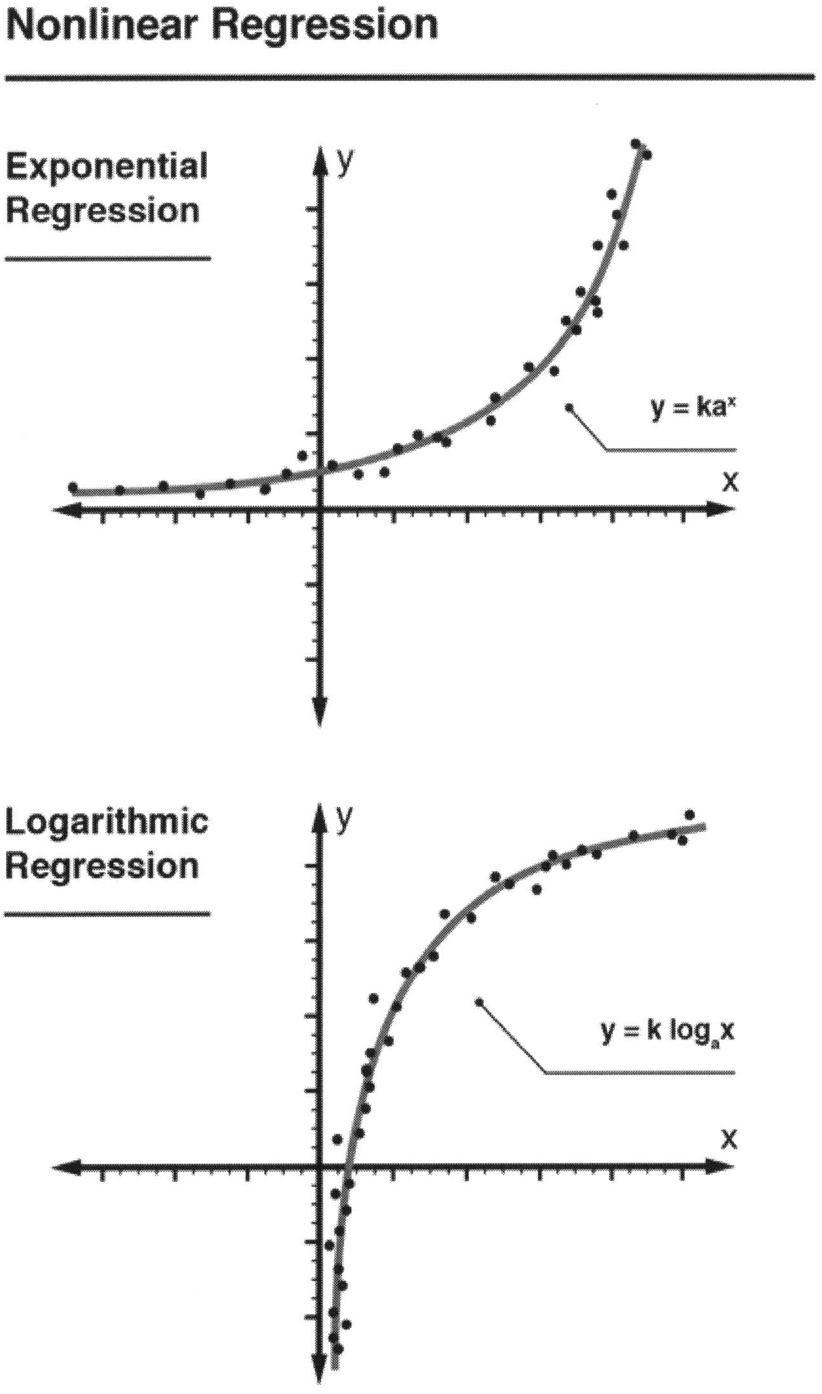


## Estimating Data Points

Regression lines can be used to estimate data points not already given. For example, if an equation of a line is found that fit the temperature and beach visitor data set, its input is the average daily temperature and its output is the projected number of visitors. Thus, the number of beach visitors on a 100-degree day can be estimated. The output is a data point on the regression line, and the number of daily visitors is expected to be greater than on a 96-degree day because the regression line has a positive slope.

## Plotting and Analyzing Residuals

Once the function is found that fits the data, its accuracy can be calculated. Therefore, how well the line fits the data can be determined. The difference between the actual dependent variable from the data set and the estimated value located on the regression line is known as a ***residual***. Therefore, the residual is known as the predicted value $\hat{y}$ minus the actual value $y$. A residual is calculated for each data point and can be plotted on the scatterplot. If all the residuals appear to be approximately the same distance from the regression line, the line is a good fit. If the residuals seem to differ greatly across the board, the line isn't a good fit.

## Interpreting the Regression Line

The formula for a regression line is $y = mx + b$, where $m$ is the slope and $b$ is the y-intercept. Both the slope and y-intercept are found in the *Method of Least Squares*, which is the process of finding the equation of the line through minimizing residuals. The slope represents the rate of change in y as $x$ gets larger. Therefore, because y is the dependent variable, the slope actually provides the predicted values given the independent variable. The y-intercept is the predicted value for when the independent variable equals zero. In the temperature example, the y-intercept is the expected number of beach visitors for a very cold average daily temperature of zero degrees.

## Correlation Coefficient

The *correlation coefficient (r)* measures the association between two variables. Its value is between -1 and 1, where -1 represents a perfect negative linear relationship, 0 represents no relationship, and 1 represents a perfect positive linear relationship. A *negative linear relationship* means that as $x$ values increase, y values decrease. A *positive linear relationship* means that as x values increase, y values increase. The formula for computing the correlation coefficient is:

$$r = \frac{n(\sum xy) - (\sum x)(\sum y)}{\sqrt{n(\sum x^2) - (\sum x)^2}\sqrt{n(\sum y^2) - (\Sigma y)^2}}$$

$n$ is the number of data points

Both Microsoft Excel® and a graphing calculator can evaluate this easily once the data points are entered. A correlation greater than 0.8 or less than -0.8 is classified as "strong" while a correlation between -0.5 and 0.5 is classified as "weak."

## Correlation Versus Causation

Correlation and causation have two different meanings. If two values are correlated, there is an association between them. However, correlation doesn't necessarily mean that one variable causes the other. *Causation* (or "cause and effect") occurs when one variable causes the other. Average daily temperature and number of beachgoers are correlated and have causation. If the temperature increases, the change in weather causes more people to go to the beach. However, alcoholism and smoking are correlated but don't have causation. The more someone drinks the more likely they are to smoke, but drinking alcohol doesn't cause someone to smoke.

## The Law of Large Numbers and the Central Limit Theorem

The *Law of Large Numbers* states that as the number of experiments increase, the actual ratio of outcomes will approach the theoretical probability. The *Central Limit Theorem* states that through using a sufficiently large sample size *N*, meaning over 30, the sampling distribution of the mean approaches a normal distribution with a mean of $\mu$ and variance of $\sigma^2/N$. The variance of the actual population is $\sigma^2$ and its mean is $\mu$. In other words, as the sample size increases, the distribution will behave normally.

### Estimating Parameters

A point estimate of a population parameter is a single statistic. For example, the sample mean is a point estimate of the population mean. Once all calculations are made, a confidence interval is used to express the accuracy of the sampling method used. The confidence interval consists of a confidence level, the statistic, and a margin of error. A 95% confidence level indicates that 95% of all confidence intervals will contain the population parameter. Also, the margin of error gives a range of values above and below the sample statistic, which helps to form a confidence interval.

### The Principles of Hypotheses Testing

The *P*-value approach to hypothesis testing involves assuming a null hypothesis is true and then determining the probability of a test statistic in the direction of the alternative hypothesis. The test statistic is defined as the *t*-statistic $t^* = \frac{\bar{x}-\mu}{s/\sqrt{n}}$, which follows a *t*-distribution with *n*-1 degrees of freedom. The *P*-value is then calculated as the probability that if the null hypothesis is true, a more extreme test statistic in the direction of the alternative hypothesis would be observed. A significance level, $\alpha$, is set (usually at 0.05 or 0.001) and the *P*-value is compared to $\alpha$. If $P \leq \alpha$, one rejects the null hypothesis and accepts the alternative hypothesis. If $P > \alpha$, one accepts the null hypothesis.

## Probabilities

### Counting Techniques

There are many counting techniques that can help solve problems involving counting possibilities. For example, the **Addition Principle** states that if there are $m$ choices from Group 1 and $n$ choices from Group 2, then $n + m$ is the total number of choices possible from Groups 1 and 2. For this to be true, the groups can't have any choices in common. The **Multiplication Principle** states that if Process 1 can be completed $n$ ways and Process 2 can be completed $m$ ways, the total number of ways to complete both Process 1 and Process 2 is $n \times m$ . For this rule to be used, both processes must be independent of each other. Counting techniques also involve permutations. A **permutation** is an arrangement of elements in a set for which order must be considered. For example, if three letters from the alphabet are chosen, ABC and BAC are two different permutations. The multiplication rule can be used to determine the total number of possibilities. If each letter can't be selected twice, the total number of possibilities is $26 \times 25 \times 24 = 15{,}600$. A formula can also be used to calculate this total. In general, the notation $P(n,r)$ represents the number of ways to arrange $r$ objects from a set of $n$ and, the formula is:

$$P(n,r) = \frac{n!}{(n-r)!}$$

In the previous example:

$$P(26,3) = \frac{26!}{23!} = 15{,}600$$

Contrasting permutations, a **combination** is an arrangement of elements in which order doesn't matter. In this case, ABC and BAC are the same combination. In the previous scenario, there are six permutations that represent each single combination. Therefore, the total number of possible combinations is $15{,}600 \div 6 = 2{,}600$. In general, $C(n,r)$ represents the total number of combinations of $n$ items selected $r$ at a time where order doesn't matter, and the formula is:

$$C(n,r) = \frac{n!}{(n-r)!\ r!}$$

Therefore, the following relationship exists between permutations and combinations:

$$C(n,r) = \frac{P(n,r)}{n!} = \frac{P(n,r)}{P(r,r)}$$

## Fundamental Counting Principle

The **fundamental counting principle** states that if there are *m* possible ways for an event to occur, and *n* possible ways for a second event to occur, there are $m \cdot n$ possible ways for both events to occur. For example, there are two events that can occur after flipping a coin and six events that can occur after rolling a die, so there are $2 \times 6 = 12$ total possible event scenarios if both are done simultaneously. This principle can be used to find probabilities involving finite sample spaces and independent trials because it calculates the total number of possible outcomes. For this principle to work, the events must be independent of each other.

## Independence and Conditional Probability

### *Sample Subsets*

A sample can be broken up into subsets that are smaller parts of the whole. For example, consider a sample population of females. The sample can be divided into smaller subsets based on the characteristics of each female. There can be a group of females with brown hair and a group of females that wear glasses. There also can be a group of females that have brown hair *and* wear glasses. This "and" relates to the **intersection** of the two separate groups of brunettes and those with glasses. Every female in that intersection group has both characteristics. Similarly, there also can be a group of females that either have brown hair *or* wear glasses. The "or" relates to the union of the two separate groups of brunettes and glasses. Every female in this group has at least one of the characteristics. Finally, the group of females who do not wear glasses can be discussed. This "not" relates to the **complement** of the glass-wearing group. No one in the complement has glasses. **Venn diagrams** are useful in highlighting these ideas. When discussing statistical experiments, this idea can also relate to events instead of characteristics.

### *Verifying Independent Events*

Two events aren't always independent. For example, females with glasses and brown hair aren't independent characteristics. There definitely can be overlap because females with brown hair can wear glasses. Also, two events that exist at the same time don't have to have a relationship. For example, even if all females in a given sample are wearing glasses, the characteristics aren't related. In this case, the probability of a brunette wearing glasses is equal to the probability of a female being a brunette multiplied by the probability of a female wearing glasses. This mathematical test of $P(A \cap B) = P(A)P(B)$ verifies that two events are independent.

### *Conditional Probability*

**Conditional probability** is the probability that event A will happen given that event B has already occurred. An example of this is calculating the probability that a person will eat dessert once they have eaten dinner. This is different than calculating the probability of a person just eating dessert. The formula for the conditional probability of event A occurring given B is $P(A|B) = \frac{P\,(A \text{ and } B)}{P(B)}$, and it's defined to be the probability of both A and B occurring divided by the probability of event B occurring. If A and B are independent, then the probability of both A and B occurring is equal to $P(A)P(B)$, so $P(A|B)$ reduces to just $P(A)$. This means that A and B have no relationship, and the probability of A occurring is the same as the conditional probability of A occurring given B. Similarly, $P(B|A) = \frac{P\,(B \text{ and } A)}{P(A)} = P(B)$ if A and B are independent.

### *Independent Versus Related Events*

To summarize, conditional probability is the probability that an event occurs given that another event has happened. If the two events are related, the probability that the second event will occur changes if the other event has happened. However, if the two events aren't related and are therefore independent, the first event to occur won't impact the probability of the second event occurring.

### *Measuring Probabilities with Two-Way Frequency Tables*

When measuring event probabilities, two-way frequency tables can be used to report the raw data and then used to calculate probabilities. If the frequency tables are translated into relative frequency tables, the probabilities presented in the table can be plugged directly into the formulas for conditional probabilities. By plugging in the correct frequencies, the data from the table can be used to determine if events are independent or dependent.

### *Differing Probabilities*

The probability that event A occurs differs from the probability that event A occurs given B. When working within a given model, it's important to note the difference. $P(A|B)$ is determined using the formula $P(A|B) = \frac{P\,(A \text{ and } B)}{P(B)}$ and represents the total number of A's outcomes left that could occur after B occurs. $P(A)$ can be calculated without any regard for B. For example, the probability of a student finding a parking spot on a busy campus is different once class is in session.

### *The Addition Rule*

The probability of event A or B occurring isn't equal to the sum of each individual probability. The probability that both events can occur at the same time must be subtracted from this total. This idea is shown in the **addition rule**:

$$P(A \text{ or } B) = P(A) + P(B) - P(A \text{ and } B)$$

The addition rule is another way to determine the probability of compound events that aren't mutually exclusive. If the events are mutually exclusive, the probability of both A and B occurring at the same time is 0.

## Computing Probabilities

### *Simple and Compound Events*

A **simple event** consists of only one outcome. The most popular simple event is flipping a coin, which results in either heads or tails. A **compound event** results in more than one outcome and consists of more than one simple event. An example of a compound event is flipping a coin while tossing a die. The result is either heads or tails on the coin and a number from one to six on the die. The probability of a simple event is calculated by dividing the number of possible outcomes by the total number of outcomes. Therefore, the probability of obtaining heads on a coin is $\frac{1}{2}$, and the probability of rolling a 6 on a die is $\frac{1}{6}$. The probability of compound events is calculated using the basic idea of the probability of simple events. If the two events are independent, the probability of one outcome is equal to the product of the probabilities of each simple event. For example, the probability of obtaining heads on a coin and rolling a 6 is equal to:

$$\frac{1}{2} \times \frac{1}{6} = \frac{1}{12}$$

The probability of either A or B occurring is equal to the sum of the probabilities minus the probability that both A and B will occur. Therefore, the probability of obtaining either heads on a coin or rolling a 6 on a die is:

$$\frac{1}{2} + \frac{1}{6} - \frac{1}{12} = \frac{7}{12}$$

The two events aren't mutually exclusive because they can happen at the same time. If two events are mutually exclusive, and the probability of both events occurring at the same time is zero, the probability of event A or B occurring equals the sum of both probabilities. An example of calculating the probability of two mutually exclusive events is determining the probability of pulling a king or a queen from a deck of cards. The two events cannot occur at the same time.

### *Uniform and Non-Uniform Probability Models*

A **uniform probability model** is one where each outcome has an equal chance of occurring, such as the probabilities of rolling each side of a die. A **non-uniform probability model** is one where each outcome has an unequal chance of occurring. In a uniform probability model, the conditional probability formulas for $P(B|A)$ and $P(A|B)$ can be multiplied by their respective denominators to obtain two formulas for $P(A \text{ and } B)$. Therefore, the multiplication rule is derived as:

$$P(A \text{ and } B) = P(A)P(B|A) = P(B)P(A|B)$$

In a model, if the probability of either individual event is known and the corresponding conditional probability is known, the multiplication rule allows the probability of the joint occurrence of A and B to be calculated.

# Practice Questions

1. Two cards are drawn from a shuffled deck of 52 cards. What's the probability that both cards are Kings if the first card isn't replaced after it's drawn?

   a. $\frac{1}{169}$

   b. $\frac{1}{221}$

   c. $\frac{1}{13}$

   d. $\frac{4}{13}$

   e. $\frac{1}{104}$

2. What's the probability of rolling a 6 at least once in two rolls of a die?

   a. $\frac{1}{3}$

   b. $\frac{1}{36}$

   c. $\frac{1}{6}$

   d. $\frac{1}{12}$

   e. $\frac{11}{36}$

3. Given the set $A = \{1, 2, 3, 4, 5, 6, 7, 8, 9, 10\}$ and $B = \{1, 2, 3, 4, 5\}$, what is $A - (A \cap B)$?

   a. $\{6, 7, 8, 9, 10\}$
   b. $\{1, 2, 3, 4, 5\}$
   c. $\{1, 2, 3, 4, 5, 6, 7, 8, 9, 10\}$
   d. $\emptyset$
   e. $\{-1, -2, -3, -4, -5\}$

4. Let $p$ = "Alex is an engineering major," $q$ = "Alex is not an English major," $r$ = "Alex's sister is a history major," $s$ = "Alex's sister has been to Germany," and t = "Alex's sister has been to Austria." Which of the following answers represents the statement "Alex is an engineering and English major, but his sister is a history major who hasn't been to either Germany or Austria."?

   a. $p \wedge {\sim}q \wedge (r \vee ({\sim}s \vee {\sim}t))$
   b. $p \wedge q \wedge r \vee ({\sim}s \wedge {\sim}t)$
   c. $p \wedge {\sim}q \wedge r \wedge ({\sim}s \vee {\sim}t)$
   d. $p \wedge q \wedge (r \vee ({\sim}s \wedge {\sim}t))$
   e. $p \wedge {\sim}q \wedge r \vee ({\sim}s \wedge {\sim}t)$

5. For a group of 20 men, the median weight is 180 pounds and the range is 30 pounds. If each man gains 10 pounds, which of the following would be true?
   a. The median weight will increase, and the range will remain the same.
   b. The median weight and range will both remain the same.
   c. The median weight will stay the same, and the range will increase.
   d. The median weight and range will both increase.
   e. The median weight will increase and the range will decrease.

6. A sample data set contains the following values: 1, 3, 5, 7. What's the standard deviation of the set?
   a. 2.58
   b. 4
   c. 6.23
   d. 1.1
   e. 0.25

7. Dwayne has received the following scores on his math tests: 78, 92, 83, and 97. What score must Dwayne get on his next math test to have an overall average of 90?
   a. 89
   b. 98
   c. 95
   d. 94
   e. 100

8. What is the probability of randomly picking the winner and runner-up from a race of 4 horses and distinguishing which is the winner?
   a. $\frac{1}{4}$
   b. $\frac{1}{2}$
   c. $\frac{1}{16}$
   d. $\frac{1}{12}$
   e. $\frac{1}{64}$

9. What is the interquartile range (IQR) of the following data set? 1, 4, 6, 6, 9, 10, 12, 17, 18
   a. 10
   b. 9
   c. 14.5
   d. 5
   e. 9.5

$$G = .035O + .26$$

10. The linear regression model above is based on an analysis of the price of a gallon of gas (G) at 15 gas stations compared to the price of a barrel of oil (O) at the time. Based on this model, which of the following statements are true?

I. There is a negative correlation between *G* and *O*.
II. When oil is $55 per barrel then gas is approximately $2.19 per gallon.
III. The slope of the line indicates that as *O* increases by 1, *G* increases by .035.
IV. If the price of oil increases by $8 per barrel then the price of gas will increase by approximately $0.18 per gallon.

a. I and II
b. II only
c. II and III
d. I and III
e. II, III, and IV

11. Five students take a test. The scores of the first four students are 80, 85, 75, and 60. If the median score is 80, which of the following could NOT be the score of the fifth student?

a. 60
b. 80
c. 85
d. 100
e. 90

12. A pair of dice is thrown, and the sum of the two scores is calculated. What's the expected value of the roll?

a. 5
b. 6
c. 7
d. 8
e. 9

13. Given the value of a given stock at monthly intervals, which graph should be used to best represent the trend of the stock?

a. Box plot
b. Line plot
c. Line graph
d. Circle graph
e. Dot plot

14. The following graph compares the various test scores of the top three students in each of these teacher's classes. Based on the graph, which teacher's students had the smallest range of test scores?

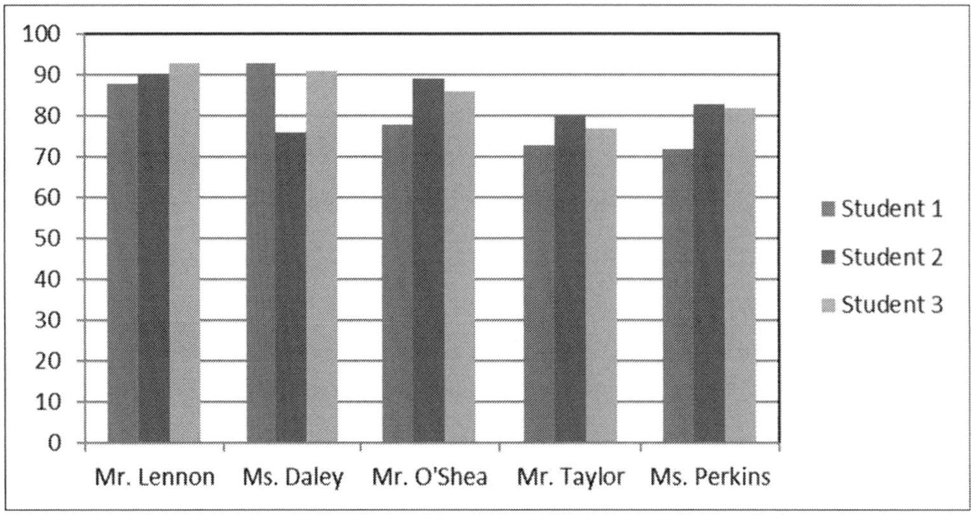


a. Mr. Lennon
b. Mr. O'Shea
c. Mr. Taylor
d. Ms. Daley
e. Ms. Perkins

# Answer Explanations

**1. B:** For the first card drawn, the probability of a King being pulled is $\frac{4}{52}$. Since this card isn't replaced, if a King is drawn first the probability of a King being drawn second is $\frac{3}{51}$. The probability of a King being drawn in both the first and second draw is the product of the two probabilities: $\frac{4}{52} \times \frac{3}{51} = \frac{12}{2652}$. This fraction, when divided by 12, equals $\frac{1}{221}$.

**2. E:** The addition rule is necessary to determine the probability because a 6 can be rolled on either roll of the die. The rule used is:

$$P(A \text{ or } B) = P(A) + P(B) - P(A \text{ and } B)$$

The probability of a 6 being individually rolled is $\frac{1}{6}$ and the probability of a 6 being rolled twice is:

$$\frac{1}{6} \times \frac{1}{6} = \frac{1}{36}$$

Therefore, the probability that a 6 is rolled at least once is:

$$\frac{1}{6} + \frac{1}{6} - \frac{1}{36} = \frac{11}{36}$$

**3. A:** $(A \cap B)$ is equal to the intersection of the two sets A and B, which is $\{1, 2, 3, 4, 5\}$. $A - (A \cap B)$ is equal to the elements of A that are *not* included in the set $(A \cap B)$. Therefore:

$$A - (A \cap B) = \{6, 7, 8, 9, 10\}$$

**4. C:** "Alex is an engineering and English major, but his sister is a history major who hasn't been to either Germany or Austria" can be rewritten as "$p$ and not $q$ and $r$ and not $s$ or not $t$." Using logical symbols, this is written as:

$$p \wedge \sim q \wedge r \wedge (\sim s \vee \sim t)$$

**5. A:** If each man gains 10 pounds, every original data point will increase by 10 pounds. Therefore, the man with the original median will still have the median value, but that value will increase by 10. The smallest value and largest value will also increase by 10 and, therefore, the difference between the two won't change. The range does not change in value and, thus, remains the same.

**6. A:** First, the sample mean must be calculated.

$$\bar{x} = \frac{1}{4}(1 + 3 + 5 + 7) = 4$$

The standard deviation of the data set is:

$$\sigma = \sqrt{\frac{\sum(x - \bar{x})^2}{n - 1}}$$

and $n = 4$ represents the number of data points.

Therefore:

$$\sigma = \sqrt{\frac{1}{3}[(1-4)^2 + (3-4)^2 + (5-4)^2 + (7-4)^2]} = \sqrt{\frac{1}{3}(9+1+1+9)} = 2.58$$

**7. E:** To find the average of a set of values, add the values together and then divide by the total number of values. In this case, include the unknown value of what Dwayne needs to score on his next test, in order to solve it.

$$\frac{78 + 92 + 83 + 97 + x}{5} = 90$$

Add the unknown value to the new average total, which is 5. Then multiply each side by 5 to simplify the equation, resulting in:

$$78 + 92 + 83 + 97 + x = 450$$

$$350 + x = 450$$

$$x = 100$$

Dwayne would need to get a perfect score of 100 in order to get an average of at least 90.

Test this answer by substituting back into the original formula.

$$\frac{78 + 92 + 83 + 97 + 100}{5} = 90$$

**8. D:** $\frac{1}{12}$. The probability of picking the winner of the race is $\frac{1}{4}$, or $\left(\frac{number\ of\ favorable\ outcomes}{number\ of\ total\ outcomes}\right)$. Assuming the winner was picked on the first selection, three horses remain from which to choose the runner-up (these are dependent events). Therefore, the probability of picking the runner-up is $\frac{1}{3}$. To determine the probability of multiple events, the probability of each event is multiplied:

$$\frac{1}{4} \times \frac{1}{3} = \frac{1}{12}$$

**9. E:** The interquartile range (IQR) is the difference between the value of the 75$^{th}$ percentile and the 25$^{th}$ percentile of a data set, or the third and first quartiles. The first step in calculating the IQR is to arrange the numbers in the data set from lowest to highest value, as they are in this problem. When the number of data points in the set is odd (as it is in this problem), the median, which is the second quartile, is not included in the upper and lower quartiles. The remaining data points are then divided into two equal groups. Then, the median of both of these groups is found, remembering that the median is the average of the two middle points if the number of values is even, as it is in this case, with four numbers. The median of the upper group is the upper quartile and the median of the lower group is the lower quartile. Finally, the IQR is calculated as the difference between the median of the upper quartile and that of the lower quartile.

So, for this problem, the data set provided was: 1, 4, 6, 6, 9, 10, 12, 17, 18.

There are nine values, so the median (9) is excluded. This makes the lower group 1, 4, 6, and 6. Then the upper group is 10, 12, 17, and 18. The median of the lower group is 5 (the average of 4 and 6) and the median of the upper group is 14.5 (the average of 12 and 17).

Thus, the IQR is $14.5 - 5 = 9.5$.

**10. C:** II and III are the only true statements. If graphed this line would have a positive correlation, which make statement I false. Also, if the price of oil increases by $8 per barrel then gas price would increase by $.035(8) = \$0.28$ per gallon. This makes statement IV false as well.

**11. A:** Lining up the given scores provides the following list: 60, 75, 80, 85, and one unknown. Because the median needs to be 80, it means 80 must be the middle data point out of these five. Therefore, the unknown data point must be the fourth or fifth data point, meaning it must be greater than or equal to 80. The only answer that fails to meet this condition is 60.

**12. C:** The expected value is equal to the total sum of each product of individual score and probability. There are 36 possible rolls. The probability of rolling a 2 is $\frac{1}{36}$. The probability of rolling a 3 is $\frac{2}{36}$. The probability of rolling a 4 is $\frac{3}{36}$. The probability of rolling a 5 is $\frac{4}{36}$. The probability of rolling a 6 is $\frac{5}{36}$. The probability of rolling a 7 is $\frac{6}{36}$. The probability of rolling an 8 is $\frac{5}{36}$. The probability of rolling a 9 is $\frac{4}{36}$. The probability of rolling a 10 is $\frac{3}{36}$. The probability of rolling an 11 is $\frac{2}{36}$. Finally, the probability of rolling a 12 is $\frac{1}{36}$.

Each possible outcome is multiplied by the probability of it occurring. Like this:

$$2 \times \frac{1}{36} = a$$

$$3 \times \frac{2}{36} = b$$

And so forth.

Then all of those results are added together:

$$a + b + c \ldots = expected\ value$$

In this case, it equals 7.

**13. C:** The scenario involves data consisting of two variables: month and stock value. Box plots display data consisting of values for one variable. Therefore, a box plot is not an appropriate choice. Both line plots (which are also called dot plots) and circle graphs are used to display frequencies within categorical data. Neither can be used for the given scenario. Line graphs display two numerical variables on a coordinate grid and show trends among the variables, so this is the correct choice.

**14. A:** To calculate the range in a set of data, subtract the lowest value from the highest value. In this graph, the range of Mr. Lennon's students is 5, which can be seen physically in the graph as having the smallest difference between the highest value and the lowest value compared with the other teachers.

Dear SAT Math 2 Test Taker,

We would like to start by thanking you for purchasing this study guide for your SAT Math 2 exam. We hope that we exceeded your expectations.

Our goal in creating this study guide was to cover all of the topics that you will see on the test. We also strove to make our practice questions as similar as possible to what you will encounter on test day. With that being said, if you found something that you feel was not up to your standards, please send us an email and let us know.

We would also like to let you know about other books in our catalog that may interest you.

## SAT Biology

This can be found on Amazon: amazon.com/dp/1628458933

## SAT Literature

amazon.com/dp/1628456280

## SAT

amazon.com/dp/1628458984

## ACCUPLACER

amazon.com/dp/162845945X

We have study guides in a wide variety of fields. If the one you are looking for isn't listed above, then try searching for it on Amazon or send us an email.

Thanks Again and Happy Testing!
Product Development Team
info@studyguideteam.com

**Photo Credits**
The following photo is licensed under CC BY 2.5 (creativecommons.org/licenses/by/2.5/)

"Black cherry tree histogram" by Mwtoews
(https://commons.wikimedia.org/wiki/Histogram#/media/File:Black_cherry_tree_histogram.svg)

# FREE Test Taking Tips DVD Offer

To help us better serve you, we have developed a Test Taking Tips DVD that we would like to give you for FREE. **This DVD covers world-class test taking tips that you can use to be even more successful when you are taking your test.**

All that we ask is that you email us your feedback about your study guide. Please let us know what you thought about it – whether that is good, bad or indifferent.

To get your **FREE Test Taking Tips DVD**, email freedvd@studyguideteam.com with "FREE DVD" in the subject line and the following information in the body of the email:

a. The title of your study guide.

b. Your product rating on a scale of 1-5, with 5 being the highest rating.

c. Your feedback about the study guide. What did you think of it?

d. Your full name and shipping address to send your free DVD.

If you have any questions or concerns, please don't hesitate to contact us at freedvd@studyguideteam.com.

Thanks again!

Made in the USA
Columbia, SC
15 October 2020

22933270R00111